Compound Semiconductors for Energy Applications and Environmental Sustainability—2011

T0301884

MATERIALS RESEARCH SOCIETY
SYMPOSIUM PROCEEDINGS VOLUME 1324

Compound Semiconductors for Energy Applications and Environmental Sustainability–2011

Symposium held April 25–29, 2011, San Francisco, California, U.S.A.

EDITORS

L. Douglas Bell

Jet Propulsion Laboratory
Pasadena, California, U.S.A.

F. (Shadi) Shahedipour-Sandvik

University at Albany–SUNY
Albany, New York, U.S.A.

Kenneth A. Jones

U.S. Army Research Laboratory
Adelphi, Maryland, U.S.A.

Daniel Schaadt

Karlsruhe Institute of Technology
Karlsruhe, Germany

Blake S. Simpkins

Naval Research Laboratory
Washington, District of Columbia, U.S.A.

Miguel A. Contreras

National Renewable Energy Laboratory
Golden, Colorado, U.S.A.

Materials Research Society
Warrendale, Pennsylvania

CAMBRIDGE
UNIVERSITY PRESS

CAMBRIDGE UNIVERSITY PRESS
Cambridge, New York, Melbourne, Madrid, Cape Town,
Singapore, São Paulo, Delhi, Tokyo, Mexico City

Cambridge University Press
32 Avenue of the Americas, New York, NY 10013-2473, USA

www.cambridge.org
Information on this title: www.cambridge.org/9781605113012

Materials Research Society
506 Keystone Drive, Warrendale, PA 15086, USA
http://www.mrs.org

© Materials Research Society 2012

First published 2012

CODEN: MRSPDH

ISBN: 978-1-60511-301-2 Hardback

CONTENTS

CdTe/CdS

CIGS/CIS

*Invited Paper

PREFACE

This volume contains a subset of Oral and Poster presentations that were made during Symposium D, "Compound Semiconductors for Energy Applications and Environmental Sustainability", at the 2011 MRS Spring Meeting held April 25-29 in San Francisco, California.

Compound semiconductors have long been an integral part of everyday life. Many of these semiconductors exhibit direct band gaps that are tailorable over a wide range of energy. This property can be leveraged for many energy-related applications such as efficient lighting, high-efficiency solar cells, and efficient switching. Recent progress on their potential as emitters, sensing devices in biological and chemical environments, and high efficiency power devices demonstrates their impact on conservation of energy and environment, and on mitigation of climate change. Compound semiconductor-based photovoltaic systems are emerging as an economical means of generating renewable energy through the use of concentrator technologies. The significant funding by various federal (e.g., Department of Energy's Sunshot Initiative) and state agencies as well as industry clearly signifies the importance of green energy and its role in supplementing and potentially replacing greenhouse-generating fuels.

Use of compound semiconductors such as the III-nitride family of materials has played the most significant role in the realization of solid state lighting as a viable means for potentially full replacement of the traditional means of lighting such as fluorescent and incandescent lighting. In order to fully realize this potential, fundamental scientific questions such as efficiency droop in green LEDs need to be addressed. Research into the use of III-nitrides for photovoltaic (PV) applications is also significant, attracting much attention as full-solar spectrum PV materials. InSb and CuInGaSe are also being researched as the major players in the PV field. Nanostructures based on these compound semiconductors show favorable properties such as more efficient collection and transport of carriers. Multijunction solar cells using InGaAs and InGaP are important players in space applications, where efficiency is the critical metric, but continued progress may make these contenders in the commercial marketplace.

These are only a few of the many examples of the significant role compound semiconductors will play in our energy future. Understanding the interaction of these compounds with their environment, including any potentially negative impact they may have on organisms and the natural environment, are other topics that need much research. This volume contains reports from internationally known experts on the state of compound semiconductor-based devices with applications to environmental conservation and energy use reduction, challenges associated with realization of such devices, and obstacles to their widespread use.

The Organizers wish to thank all who contributed to the success of Symposium D, in particular the authors, reviewers, and the MRS staff.

L. Douglas Bell
F. (Shadi) Shahedipour-Sandvik
Kenneth. A. Jones
Daniel Schaadt
Blake S. Simpkins
Miguel A. Contreras

September 2011

MATERIALS RESEARCH SOCIETY SYMPOSIUM PROCEEDINGS

MATERIALS RESEARCH SOCIETY SYMPOSIUM PROCEEDINGS

Prior Materials Research Society Symposium Proceedings available by contacting Materials Research Society

Nitrides

Mater. Res. Soc. Symp. Proc. Vol. 1324 © 2011 Materials Research Society
DOI: 10.1557/opl.2011.960

Indium Gallium Nitride on Germanium by Molecular Beam Epitaxy

R.R. Lieten[1,2] , W.-J. Tseng[1,2], M. Leys[2], J.-P. Locquet[1], J. Dekoster[2]
[1] Department of Physics and Astronomy, K.U. Leuven, 3001 Leuven, Belgium
[2] IMEC, 3001 Leuven, Belgium

ABSTRACT

Indium containing III-Nitride layers are predominantly grown by heteroepitaxy on foreign substrates, most often Al_2O_3, SiC and Si. We have investigated the epitaxial growth of $In_xGa_{1-x}N$ (InGaN) alloys on Ge substrates. First we looked at the influence of buffer layers between the InGaN and Ge substrate. When applying a high temperature (850 °C) GaN buffer, the InGaN showed superior crystal quality. Furthermore the influence of growth parameters on the structural quality and composition of InGaN layers has been looked into. For a fixed gallium and nitrogen supply, the indium beam flux was increased incrementally. For both nitrogen- as well as for metal (Ga + In) rich growth conditions, the In incorporation increases for increasing In flux. However, for metal rich growth conditions, segregation of metallic In is observed. An optimum in crystal quality is obtained for a metal:nitrogen flux ratio close to unity. The XRD FWHM of the GaN (0002) reflection increases significantly after InGaN growth. Apparently the presence of indium deteriorates the GaN buffer during InGaN growth. The mechanism of the effect is not known yet.

INTRODUCTION

Single crystalline indium containing III-Nitrides show interesting electrical and optical properties. The ability to tune the direct band gap by changing the alloy composition is promising for solar matched photovoltaics. Indium containing III-Nitride layers are predominantly grown by heteroepitaxy on foreign substrates, most often Al_2O_3, SiC and Si [1,2,3]. This material system is promising for photovoltaic applications [2,3,4,5].

Better understanding of the growth processes is needed to improve the structural and optoelectronic properties. In this work we investigate the heteroepitaxial growth of Indium Gallium Nitride layers on germanium substrates with molecular beam epitaxy (MBE).

Previously we have demonstrated heteroepitaxial growth of GaN on Ge(111) by plasma assisted molecular beam epitaxy (PAMBE) [6]. The use of Ge substrates for III-Nitrides growth, and in particular InGaN, could be advantageous for devices in which vertical conduction is required [7]. A direct photo electrolysis cell, using photocurrent to split H_2O into H_2 and O_2, is an example of a device, which would benefit from vertical conduction. Using a back contact can much simplify such a design. Another promising application of InGaN on conducting Ge substrates is a high-efficiency solar cell. InGaN can absorb the UV and visible part of the solar spectrum and germanium the infrared part.

In this work we report on the influence of buffer layers and growth parameters on the structural quality and composition of InGaN layers grown on Germanium (111) substrates. X-ray

3

diffraction (XRD) has been used to determine the indium content, crystal quality, the homogeneity and the presence of InN and metallic indium clusters. The surface morphology and layer thickness has been investigated by atomic force microscopy (AFM) and scanning electron microscopy (SEM).

EXPERIMENTAL DETAILS

Germanium substrates with (111) orientation and diameter of 2 inch were chemically cleaned to remove metallic contamination, particles and native oxide from the surface, just before loading into the MBE system. Subsequently, the samples were degassed at 450 °C in vacuum with a background pressure of 1×10^{-9} Torr. The cleanliness of the surface was confirmed by reflection high-energy electron diffraction (RHEED), which showed a reconstructed surface. Substrate temperatures were measured by thermocouple. A N_2 flow of 1.2 SCCM and a radio frequency power of 250 W have been used. These settings correspond to an atomic nitrogen beam equivalent pressure of around 4.5×10^{-7} Torr or 8.0×10^{14} atoms cm^{-2} s^{-1}. A few monolayers of single crystalline Ge_3N_4 are formed by exposing the substrate to nitrogen plasma at 850 °C [8]. The presence of this crystalline Ge_3N_4 layer just before the start of GaN growth, has been used to explain the epitaxial growth of GaN on Ge(111) [8]. The indium content has been deduced from the position of the XRD (0002) InGaN reflection, assuming 100 % relaxation. This assumption is reasonable regarding InGaN thicknesses of around 1 micron and lattice mismatches with respect to the underlying GaN from 1.1 to 2.1 %.

DISCUSSION

Influence of buffer layer

First, we have investigated the influence of buffer layers on subsequent InGaN growth: high temperature (HT) GaN (sample A), HT GaN followed by low temperature (LT) GaN (sample B), HT GaN followed by graded InGaN (sample C), graded InGaN (sample D), and finally without any buffer layer (sample E). Following these buffer layers, InGaN is always grown for 90 minutes with a Ga beam equivalent pressure of 3.8×10^{-7} Torr. The In beam equivalent pressure is kept at 1.1×10^{-7} Torr or is increased in time to 1.1×10^{-7} Torr for the graded layers. The high temperature GaN buffer layer is grown at 850 °C by supplying a Ga flux with 1.05×10^{-6} Torr beam equivalent pressure, just after formation of Ge_3N_4. Deposition of GaN at the above mentioned growth parameters leads to the suppression of domain formation in the GaN layer by enhancing step flow growth with respect to 2D nucleation [9]. Substrate temperatures below 850 °C, for the given nitrogen plasma settings, lead to the formation of rotated GaN domains [9]. For this raison, we did not use LT GaN as starting buffer layer. The HT GaN layer thickness is 50 nm (sample A). The second buffer (sample B) that has been investigated is similar to the high temperature GaN, except that a second GaN layer is grown at low temperature. The low temperature GaN buffer has been grown at 450 °C just after HT GaN epitaxy, and is 10 nm thick. As third option a graded InGaN has been grown on a HT GaN buffer (sample C). The graded layer is grown at 450 °C and starts with Ga and In beam fluxes of 4.4×10^{-7} Torr and 0 Torr, respectively. During 10 minutes the Ga and In beam fluxes are gradually reduced to 3.8×10^{-7} and increased to 1.1×10^{-7} Torr, respectively. The graded InGaN layer has a thickness of about 100 nm. The fourth buffer was a graded InGaN layer grown directly after Ge3N4 formation (Sample D)."

As reference, InGaN has been grown without buffer layer at 450 °C directly after Ge_3N_4 formation at 850 °C (sample E). The properties of these InGaN layers, grown with different buffer layers (sample A to E) are summarized in table I.

Table I. Properties of InGaN layers grown with different buffer layers on germanium substrates. The InGaN layers are grown with the same growth parameters (Ga flux, In flux, growth temperature, growth duration).

Sample	Buffer	In conc.	Roughness	XRD ω (0002)FWHM	
				GaN	InGaN
		(%)	(nm)	(arc sec)	(arc sec)
A	HT GaN	12.4	44	1200	3686
B	HT + LT GaN	10.2	41	1158	3323
C	HT GaN + graded InGaN	12.5	65	1221	14009
D	graded InGaN	16.4	36	9540	7344
E	/	13.0	28	/	7091

XRD ω-2θscans, measured with a Panalytical X'Pert system, are shown in Fig. 1 (a). From the peaks at 15.7 ° it follows that some InN is present in the InGaN layer or on the InGaN surface. This is notably the case for the HT +LT GaN buffer (sample B) and the InGaN without buffer (sample E). Apparently these conditions lead to In segregation and InN formation. The InGaN and GaN diffraction peaks are situated around 17 ° and 17.3 °, respectively. The layers with graded InGaN buffer (samples C and D) showed as expected a broader InGaN peak, indicating a broader composition variation. When comparing InGaN with HT GaN (sample A) and HT +LT GaN (sample B) we observe a side feature when no LT GaN is used, but a less intense InN peak. This indicates that at the onset of growth more In is incorporated in the InGaN layer and less In segregates for the HT GaN buffer. The InGaN surface roughness is very high and comparable for both samples: more than 40 nm root mean square (RMS) roughness was measured by AFM. A difference in stress in the GaN buffer, could explain these differences.

When comparing the XRD rocking curves (ω scan) of the (0002) reflections, Fig. 1 (b), it is clear that the HT GaN and the HT + LT GaN buffers give the best InGaN quality. The InGaN on the HT GaN buffer showed a lower InN peak intensity. For this reason we use this buffer layer further in this work. A thickness of 50 nm has been chosen. This thickness allows measuring the GaN buffer quality with XRD. Thinner GaN buffers can be useful to preserve electrical conduction between the InGaN layer and Ge substrate.

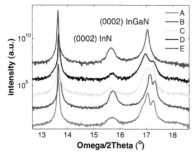 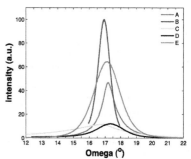

Figure 1:(a) XRD ω-2θscans for InGaN layers grown with different buffer layers on germanium. Scans are vertically offset to avoid overlap. (b) XRD ωscans (rocking curve) of InGaN (0002) reflection for InGaN grown with different buffer layers on germanium.

Influence of indium flux

We have investigated the influence of indium flux on the structural quality and indium content of InGaN layers grown on top of the optimized HT GaN buffer. The Ga flux and growth time were kept fixed at a beam equivalent pressure of 3.8 x 10^{-7} Torr and 90 minutes, respectively. The In beam flux was varied from 0.8 to 1.8 x 10^{-7} Torr. The properties of these samples (A and F to J) are summarized in table II.

Table II. Properties of InGaN layers grown with different indium fluxes on germanium substrates with 50 nm HT GaN buffer.

Sample	In flux	In conc.	Thickness	Roughness	XRD ω (0002) FWHM	
					GaN	InGaN
	(1 x 10^{-7} Torr)	(%)	(nm)	(nm)	(arc sec)	(arc sec)
F	0.8	9.7	1076	55	1108	3273
G	1.0	11.7	1105	59	1123	3901
A	1.1	12.4	1020	44	1200	3686
H	1.2	13.4	1172	39	1517	4989
I	1.6	15.2	956	73	1637	3087
J	1.8	16.9	1280	75	1989	2438

For both nitrogen (F) as well as metal (Ga+In) rich (H to J) growth conditions the indium incorporation increases linearly for increasing indium flux, as shown in Fig. 2 (b). However, for metal rich growth conditions segregation of metallic indium occurs as evident by the presence of a diffraction peak at 16.4 °. An optimum in X-ray diffraction intensity is obtained for a Ga+In to N flux ratio close to unity, with indium beam equivalent pressure of 1.0-1.1 x 10^{-7} Torr.

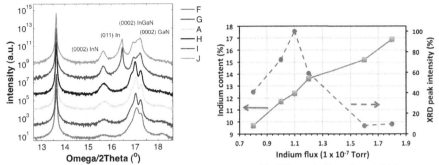

Figure 2: (a) XRD ω-2θscans for InGaN layers grown with different indium fluxes. (b) In content measured from XRD and XRD peak intensity in function of indium flux.

The thickness and root mean square (RMS) roughness have been measured by SEM and AFM, respectively. A surface area of 1.0 by 1.0 micron was scanned by AFM. All layers are quite rough, see Fig. 3 (a) and (b).

Figure 3: (a) SEM images of sample A, showing columnar InGaN. (b) AFM measurement of sample A, showing a rough surface.

The XRD FWHM of the GaN (0002) reflection increases significantly after InGaN growth. Without InGaN layer a FWHM of the (0002) reflection of 410 arc sec is obtained. When InGaN is grown on the GaN buffer, a FWHM of 1200 arc sec is measured. Apparently the GaN layer deteriorates during subsequent InGaN growth. The mechanism of the effect is not known at this time. Further investigation by transmission electron microscopy (TEM) would be interesting.

CONCLUSIONS

InGaN layers have been grown by molecular beam epitaxy on various substrates. When using Ge(111) substrates, a 50 nm thin GaN buffer layer grown at high temperature seems the best option. The influence of indium flux on the structural quality, surface morphology and indium content has been investigated. For both nitrogen as well as metal (Ga+In) rich growth conditions the indium incorporation increases for increasing indium flux. However, for metal rich growth conditions segregation of metallic indium is observed. An optimum in crystal quality is obtained for a Ga+In to N flux ratio close to unity, with an indium beam equivalent pressure of 1.0-1.1×10^{-7} Torr. The XRD FWHM of the GaN (0002) reflection increases significantly after InGaN growth. Apparently the presence of indium deteriorates the GaN buffer during InGaN growth. The mechanism of the effect is not known yet. Intermixing of the InGaN layer and GaN buffer, trough the presence of defects, is a possibility.

ACKNOWLEDGMENTS

R.R.L acknowledges support as Research Fellow of the Research Foundation - Flanders (FWO). Willem van de Graaf and Mohan Paladugu are acknowledged for assistance in molecular beam epitaxy and X-ray diffraction. This project received financial support from Interreg.

REFERENCES

1. H. J. Kim, Y. Shin, S.-Y. Kwon, H. J. Kim, S. Choi, S. Hong, C. S. Kim, J.-W. Yoon, H. Cheong, E. Yoon,Journal of Crystal Growth 310, 3004 (2008).
2. E. Matioli, C. Neufeld, M. Iza, S. C. Cruz, A.A. Al-Heji, X. Chen, R.M. Farrell, S. Keller, S. DenBaars, U. Mishra, S. Nakamura, J. Speck, and C. Weisbuch, Appl. Phys. Lett. 98, 021102 (2011).
3. L.A. Reichertz, I. Gherasoiu, K.M. Yu, V.M. Kao, W. Walukiewicz and J.W. Ager, Appl. Phys. Express 2, 122202 (2009).
4. J. R. Lang, C. J. Neufeld, C. A. Hurni, S. C. Cruz, E. Matioli, U. K. Mishra, and J. S. Speck, Appl. Phys. Lett. 98, 131115 (2011).
5. X.-M. Cai, S.-W. Zeng, and B.-P. Zhang, Applied Physics Letters 95 173504 (2009).
6. R. R. Lieten, S. Degroote, K. Cheng, M. Leys, M. Kuijk, and G. Borgh, Applied Physics Letters 89, 252118 (2006).
7. E. Trybus, G. Namkoong, W. Henderson, W.A. Doolittle, R. Liu, J. Mei, F. Ponce, M. Cheung, F. Chen, M. Furis, A. Cartwright, J. Crystal Growth 279, 311 (2005).
8. R. R. Lieten, S. Degroote, M. Kuijk, and G. Borghs, Applied Physics Letters 91, 222110 (2007).
9. R. R. Lieten, S. Degroote, M. Leys, and G. Borghs, Journal of Crystal Growth 311, 1306 (2009).
10. R.R. Lieten, O. Richard, S. Degroote, M. Leys, H. Bender, and G. Borghs, Journal of Crystal Growth 314, 71 (2011).

Mater. Res. Soc. Symp. Proc. Vol. 1324 © 2011 Materials Research Society
DOI: 10.1557/opl.2011.961

Crack-free III-nitride structures (> 3.5 μm) on silicon

Mihir Tungare[1], Jeffrey M. Leathersich[1], Neeraj Tripathi[1], Puneet Suvarna[1], Fatemeh (Shadi) Shahedipour-Sandvik[1], Timothy A. Walsh[2], Randy P. Tompkins[2] and Kenneth A. Jones[2]
[1]College of Nanoscale Science and Engineering, University at Albany, SUNY, Albany, NY 12203, U.S.A.
[2]U.S. Army Research Laboratory Sensors and Electron Devices Directorate, Adelphi, MD 20783, U.S.A.

ABSTRACT

III-nitride structures on Si are of great technological importance due to the availability of large area, epi ready Si substrates and the ability to heterointegrate with mature silicon micro and nanoelectronics. High voltage, high power density, and high frequency attributes of GaN make the III-N on Si platform the most promising technology for next-generation power devices. However, the large lattice and thermal mismatch between GaN and Si (111) introduces a large density of dislocations and cracks in the epilayer. Cracking occurs along three equivalent {1−100} planes which limits the useable device area. Hence, efforts to obtain crack-free GaN on Si have been put forth with the most commonly reported technique being the insertion of low temperature (LT) AlN interlayers. However, these layers tend to further degrade the quality of the devices due to the poor quality of films grown at a lower temperature using metal organic chemical vapor deposition (MOCVD). Our substrate engineering technique shows a considerable improvement in the quality of 2 μm thick GaN on Si (111), with a simultaneous decrease in dislocations and cracks. Dislocation reduction by an order of magnitude and crack separation of > 1 mm has been achieved. Here we combine our method with step-graded AlGaN layers and LT AlN interlayers to obtain crack-free structures greater than 3.5 μm on 2" Si (111) substrates. A comparison of these film stacks before and after substrate engineering is done using atomic force microscopy (AFM) and optical microscopy. High electron mobility transistor (HEMT) devices developed on a systematic set of samples are tested to understand the effects of our technique in combination with crack reduction techniques. Although there is degradation in the quality upon the insertion of LT AlN interlayers, this degradation is less prominent in the stack grown on the engineered substrates. Also, this methodology enables a crack-free surface with the capability of growing thicker layers.

INTRODUCTION

Growth of nitrides on silicon has gained tremendous interest over the past decade. The large scale availability, low cost, and high quality of the Si substrates combined with the large bandgap, high breakdown strength, high maximum oscillation frequency, superior noise factor, and high current attributes of GaN, makes this platform very attractive for next generation devices. The major roadblock with realizing this is the large difference in thermal expansion coefficients ($\alpha_{Si} = 2.59 \times 10^{-6}$ K^{-1} and $\alpha_{GaN} = 5.59 \times 10^{-6}$ K^{-1}) and lattice constants ($a_{Si(111)} = 3.84$ Å and $a_{GaN} = 3.189$ Å) between the two material systems [1, 2]. This leads to a large density of cracks in the epilayer along three equivalent {1−100} planes and dislocations around 10^{10} cm^{-2}

[3, 4]. Strain management techniques have been incorporated for the reduction and elimination of cracking over a large area, with the commonly used methods being LT AlN interlayers [5], AlN/GaN superlattices [6] or AlGaN/GaN superlattices [7], and graded AlGaN buffers [8]. However, a simultaneous reduction in dislocations in conjunction with the above mentioned methods has still remained a challenge.

We have developed a technique to modify the Si substrate by implantation of nitrogen ions through a high temperature (HT) AlN wetting layer into the substrate. This method has demonstrated a simultaneous and considerable reduction in cracking and dislocation defects. Crack separation of greater than 1 mm and dislocation reduction down to 10^8 cm^{-2} has been achieved in a 2μm thick GaN grown on Si [9, 10]. Here we combine our method with commonly used strain reduction techniques to obtain crack-free structures over a 2" wafer. The purpose of this study is to take advantage of the improved GaN quality resulting from the substrate engineering technique and also create thicker crack-free layers. This is necessary to reduce vertical leakage and also provide greater electrical isolation from the substrate, required for devices like HEMTs. Various AlGaN/GaN HEMT devices are developed and the impact of combining our substrate engineering technique with other crack reducing methods is explored.

EXPERIMENT

Deposition of AlN buffers on 2" Si (111) substrates is done at a temperature of about 1010 °C using a Veeco D180 MOCVD (metal organic chemical vapor deposition) reactor. The details of sample preparation prior to growth are listed elsewhere [11]. Some of these samples undergo our substrate engineering process before regrowth. This involves the implantation of nitrogen at energy of 65 keV and a dose of 2×10^{16} cm^{-2} through the AlN/Si stack to create an amorphized region within Si beneath the AlN-Si interface.

To develop crack-free structures greater than 3.5 μm, engineered AlN/Si substrate and un-treated AlN/Si substrates are used. Both types of samples underwent *in-situ* annealing for 30 minutes in N$_2$ prior to the growth of a step-graded Al$_x$Ga$_{1-x}$N buffer layer. The step-graded buffer consisted of decreasing Al percentages of 75%, 66%, 50%, 33%, and 20%, with each Al$_x$Ga$_{1-x}$N layer 100 nm thick. 1 μm GaN is grown on top of this buffer followed by three LT AlN interlayers (each 15 nm thick) separated by 250 nm of GaN. This is then followed by growth of 1.5 μm of GaN.

Four HEMT devices are fabricated on Si: (1) Sample A: without LT AlN interlayers, (2) Sample B: with LT AlN interlayers, (3) Sample C: engineered substrate without LT AlN interlayers, and (4) Sample D: engineered substrate with LT AlN interlayers. Figure 1 shows a schematic of the HEMT device.

On-wafer characterizations including optical microscopy, AFM, and low temperature (LT) photoluminescence (PL) are performed. LT PL measurements are performed at 8 K using a He-Cd 325 nm laser source. Above mentioned thick 3.5 μm structures are compared to our standard 2 μm of GaN on engineered substrate that does not have a step-graded AlGaN buffer and LT AlN interlayers. Moreover, HEMT on Si structures were processed/fabricated and measured and are compared amongst each other and also with a similar HEMT structure grown on sapphire.

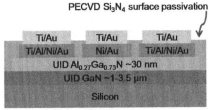

PECVD Si$_3$N$_4$ surface passivation

AlGaN/GaN HEMT structure

Figure 1 Schematic of AlGaN/GaN HEMT device structure

RESULTS AND DISCUSSION

Surface characterization of 3.5 μm structures

Optical and AFM images of these structures are shown in Figure 2. Structures on both unimplanted AlN/Si and implanted AlN/Si substrates are crack-free across the entire 2" wafer with some cracks visualized towards the periphery of the wafer. Smoother morphology seen for GaN on engineered substrate as illustrated in the 10×10 μm^2 AFM scan, with an RMS roughness of 1.6 nm as opposed to 2.4 nm for the sample that did not go through the implantation process.

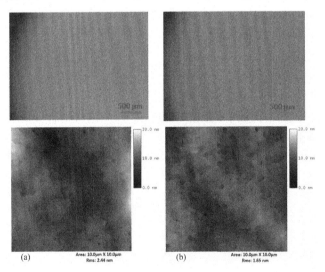

Figure 2 (a) GaN on unimplanted AlN/Si substrates (b) GaN on implanted AlN/Si substrates

11

Electrical characteristics: AlGaN/GaN HEMT

'Sample A' showed no gate control of drain current and was inconclusive towards the presence of two-dimensional electron gas (2DEG). This sample did not undergo any substrate engineering nor did it have any LT AlN interlayers. AFM of the surface of this structure showed microcracks which explain the results observed. The rest of the samples exhibited 2DEG in the low to mid 10^{12} cm^{-2}. 'Sample C' was a similar device as 'Sample A' but underwent substrate engineering. This showed better characteristics with regard to the presence of 2DEG and ability to measure field effect mobility, when compared to 'Sample A'. Presence of 2DEG and ability to measure a field effect mobility for 'Sample C', though low (average ~97 cm^2/Vs), are indications that the substrate engineering technique successfully improves the quality of the layers in terms of defects and cracks. 'Sample D' showed improved characteristics compared to 'Sample C' perhaps due to the complete absence of any cracks with the average field effect mobility improved to 135 cm^2/Vs and 2DEG concentration (n_s) of $(1.51 \pm 0.19) \times 10^{12}$ cm^{-2}. All devices showed a better pinch-off but had a high gate leakage, lower transconductance and maximum drain current, when compared to similar devices on sapphire. The maximum 2DEG obtained for the HEMT devices on Si is ~5×10^{12} cm^{-2} with an average mobility of 1041 cm^2/Vs. A threshold voltage as high as -4.38 $\pm$ 0.54 V and an average field effect mobility as high as 165 cm^2/Vs have been achieved. It is noteworthy that no attempt was made to optimize the HEMT characteristics of the devices on Si, and it was only used for the purpose of comparing various defect and crack reduction methods. Table I summarizes the electrical results for the HEMT on Si samples.

Table I. Electrical characteristics of HEMT on Si devices

	Sample B	Sample C	Sample D
Substrate	Silicon	Silicon	Silicon
n_s (cm^{-2})	$(4.82 \pm 1.02) \times 10^{12}$	$(2.56 \pm 0.94) \times 10^{11}$	$(1.51 \pm 0.19) \times 10^{12}$
Threshold voltage (V)	-4.38 $\pm$ 0.54	-2.97 $\pm$ 0.07	-2.06 $\pm$ 0.81
Avg. Field effect mobility (cm^2/Vs)	165.24	96.49	134.99
Avg. Mobility (cm^2/Vs)	1041.2		1223.7

Optical characteristics: LT PL

Figure 3 and Figure 4 show comparison between various samples with PL bands observed for each HEMT sample labeled alongside each curve. Figure 3 (a) shows a comparison between 'Sample C' and HEMT on sapphire. The optical quality of HEMT on sapphire is better due to the presence of the donor bound exciton (DBE) peak with much higher intensity (not shown here). Also, PL band at 3.45-3.46 eV has previously been associated with inversion domains [12] and no PL band in this range is observed for either sample. However, this band is observed, as shown in Figure 3b, for a sample of HEMT grown on Si with interlayers without incorporating our substrate engineering technique. The PL band centered around 3.437 eV correlates with 2DEG-h interaction, previously reported in the literature [13]. Though expected, the 2DEG-h band is not present for the HEMT on sapphire. 'Sample B' exhibits poorer quality with PL band at 3.45 eV indicating presence of inversion domains. This is a clear indication that the LT AlN interlayers are of lower quality and/or add to overall degradation of the overgrown GaN. A comparison of samples "C" and "D" shows two major differences in optical characteristics: presence of DAP peak in 'Sample D' and higher overall radiative recombination efficiency for this sample, as seen in Figure 4 (a). This indicates that inclusion of interlayers technique to our substrate ion implanted method improves the overall optical quality of the sample. Shown in Figure 4(b) is a comparison between PL characteristics of undoped GaN (uGaN) layers grown on engineered substrates with and without interlayers and AlGaN step grading. The uGaN sample on engineered substrate has lower defect assisted recombination centers with presence of DBE. In addition, this sample showed lower blue luminescence (BL) defect band centered at 2.8 eV (not shown here as data is normalized) in comparison with step-graded AlGaN and LT AlN interlayers [12].

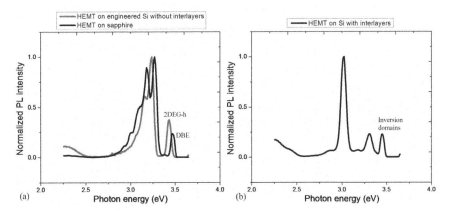

Figure 3 (a) Comparison of LT PL data for HEMT on engineered sample (Sample C) with that for HEMT on sapphire (b) LT PL for Sample B

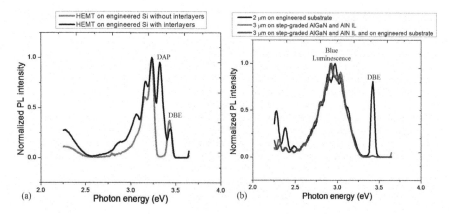

Figure 4 (a) Comparison between the LT PL data for Sample C and Sample D **(b)** LT PL comparison between 2 μm of GaN on engineered substrate with 3 μm of crack-free GaN on engineered and non-engineered substrates in conjunction with step-graded AlGaN and AlN interlayers

CONCLUSIONS

Large area dislocation and crack reduction (1 mm × 1-3 mm) is achieved simultaneously using the ion-implantation assisted substrate engineering technique. We have been able to get crack-free structures > 3.5 μm across an entire 2" wafer combining our method with conventional strain management techniques. The substrate engineering technique partially isolates the bulk of the Si substrate from the overgrown layers, both mechanically and crystallographically. Although the quality of the overgrown GaN layer is shown to be superior using our substrate engineering method, crack elimination across areas larger than 1 mm × (1-3) mm for layers thicker than 2 μm is still being developed. A combination of modified Si substrate and traditional crack eliminating methods is hence used to achieve thicker layers appropriate for electronic device structures across a large area. These structures show an improvement in the surface roughness on engineered substrates. However, addition of LT AlN interlayers seems to degrade the GaN quality and further optimization of these layers is needed to fully realize the potential of this combined method. HEMT devices on uGaN with interlayers (with and without substrate engineering) do not indicate any noticeable difference in electrical characteristics, potentially due to much degradation from LT AlN interlayers masking any benefit gained from our substrate engineering method. The influence of these interlayers on the epilayer quality seems to have a greater impact than the implantation mechanism. PL comparison of the device structures and uGaN layers on engineered Si, with and without interlayers, indicates that samples with interlayers contain inversion domains and also higher defect assisted radiative recombination. Detailed PL study is needed to ascertain the origin of some of the unidentified PL peaks.

ACKNOWLEDGMENTS

This work is partially supported by the National Science Foundation (NSF) under Award number 0904929.

REFERENCES

1. Y. Okada and Y. Tokumaru, J. Appl. Phys. **56** 314 (1984).
2. H. P. Maruska and J. J. Tietjen, Appl. Phys. Lett. **15** 327 (1969).
3. R. Jothilingam, M. W. Koch, J. B. Posthill, and G. W. Wicks, J. Electron. Mater. **30** 821 (2001).
4. A. Krost and A. Dadgar, Mater. Sci. Eng., **B93** 77 (2002).
5. A. Reiher, J. Blasing, A. Dadgar, A. Diez, and A. Krost, J. Cryst. Growth, **248** 563 (2003).
6. E. Feltin, B. Beaumont, M. Laugt, P. d. Mierry, P. Vennegues, H. Lahreche, M. Leroux, and P. Gibart, Appl. Phys. Lett., **79**, 3230 (2001).
7. S. Jang and C. Lee, J. Cryst. Growth **253**, 64 (2003).
8. A. Able, W. Wegscheider, K. Engl, and J. Zweck, J. Cryst. Growth **276**, 415 (2005).
9. M. Jamil, J. R. Grandusky, V. Jindal, F. Shahedipour-Sandvik, S. Guha, and M. Arif, Appl. Phys. Lett. **87**, 82103 (2005).
10. M. Jamil, J. R. Grandusky, V. Jindal, N. Tripathi, and F. Shahedipour-Sandvik, J. Appl. Phys. **102**, 023701 (2007).
11. M. Tungare, V. K. Kamineni, F. Shahedipour-Sandvik, and A. C. Diebold, Thin Solid Films **519**, 2929 (2011).
12. M. A. Reshchikov and H. Morkoc, J. Appl. Phys. **97**, 061301 (2005).
13. D. GuoJian, G. LiWei, X. ZhiGang, C. Yao, X. PeiQiang, J. HaiQiang, Z. JunMing, and C. Hong, Sci. China Phys. Mech. Astron. **53**, 49 (2010).

Mater. Res. Soc. Symp. Proc. Vol. 1324 © 2011 Materials Research Society
DOI: 10.1557/opl.2011.839

Comparison of Aluminum Nitride Nanowire Growth with and without Catalysts via Chemical Vapor Deposition

Kasif Teker[1] and Joseph A. Oxenham[1]
[1] Department of Physics and Engineering, Frostburg State University, 101 Braddock Road, Frostburg, MD 21532, U.S.A.

ABSTRACT

This paper presents a systematic investigation of AlN nanowire synthesis by chemical vapor deposition using Al and NH_3 on SiO_2/Si substrate and direct nitridation of mixture of Al-Al_2O_3 by NH_3. A wide variety of catalyst materials, in both discrete nanoparticle and thin film forms, have been used (Co, Au, Ni, and Fe). The growth runs have been carried out at temperatures between 800 and 1100°C mainly under H_2 as carrier gas. It was found that the most efficient catalyst in terms of nanowire formation yield was 20-nm Ni film. The AlN nanowire diameters are about 20-30 nm, about the same thickness as the Ni-film. Further studies of direct nitridation of mixture of Al-Al_2O_3 by NH_3 have resulted in high density one-dimensional nanostructure networks at 1100°C. It was observed that catalyst-free nanostructures resulted from the direct nitridation were significantly longer than that with catalysts. The analysis of the grown nanowires has been carried out by scanning electron microscopy, transmission electron microscopy, atomic force microscopy, and x-ray diffraction.

INTRODUCTION

Aluminum nitride (AlN) is a very important semiconductor material for electronic and optoelectronic applications. As a wide band gap (6.2 eV) III-V semiconductor, AlN has attracted great interest due to its inherent superior properties such as excellent thermal conductivity, low thermal expansion coefficient, high chemical stability, high electrical resistivity, and low electron affinity [1-3]. One-dimensional aluminum nitride (AlN) nanostructures are important not only for understanding fundamental concepts underlying the observed electronic, optical, and mechanical properties of materials, but also for potential applications in several fields including power transistors, heat sinks, surface acoustic wave filters, resonators, sensors, and piezoelectric nanogenerators. Thus, significant research has been devoted to the synthesis of 1-D AlN nanostructures with various fabrication methods. These fabrication methods include metal organic vapor deposition (MOCVD) [4], arc discharging process [5], chloride-assisted growth [6], carbothermal reduction [7], gas reduction nitridation [8], and CVD [9-15]. In addition, group III-nitride nanostructures attract interest due to their significant piezoelectric effect, which provides great potential for the integration of nanoelectronics and piezoelectricity [16].

This paper presents a systematic investigation of 1-D AlN nanostructure synthesis by chemical vapor deposition using Al and NH_3 on SiO_2/Si substrate and direct nitridation of mixture of Al-Al_2O_3 by NH_3 as source materials. A wide variety of catalyst materials, in both discrete nanoparticle and thin film forms, have been used (Co, Au, Ni, and Fe). Furthermore, we have conducted tests about catalyst-free growth of AlN nanostructures through direct nitridation of mixture of Al-Al_2O_3 by NH_3.

EXPERIMENTAL DETAILS

AlN nanowire growth has taken place in a resistively heated hot-wall 25-mm horizontal LPCVD reactor. Si and SiO_2/Si substrates were used for the catalyst assisted growth. 400 nm thick SiO_2 was formed by thermal oxidation of Si. Various catalyst materials including Ni (20 nm size nanoparticle and thin film), Au of 20 nm, Fe of 40 nm, and Co of 25 nm were used. All the catalyst materials have been placed on SiO_2/Si substrate except Ni film, which was pre-deposited on Si substrate by sputtering. The substrate was ultrasonically cleaned in acetone, isopropyl alcohol, de-ionized water and dried with nitrogen. Nanoparticle solution was applied to the substrate surface and dried. An alumina boat containing both the substrate and Al (99.97 % purity, about 30 mg) was loaded into the CVD reactor. A mixture of Al and Al_2O_3 (1: 1) source materials was used for the direct nitridation studies. Moreover, the mixture of Al and Al_2O_3 was placed into alumina boat and the growth took place directly on top of the source mixture. Following loading, the reactor was evacuated and purged three times with hydrogen (99.999 %). After purging cycles, the reactor was heated to the growth temperature (typically between 800 and 1100°C) under hydrogen. Then, the growth was carried out by flowing NH_3 (99.99%) and H_2 gases through the reactor for typically about 15 min. The gas flow rates were controlled by mass flow controllers and set to 300 sccm for both H_2 and NH_3. After the growth, NH_3 was shut off and the reactor cooled down under H_2 flow until 250°C. Then, the furnace naturally cooled down to room temperature.

The samples have been characterized by scanning electron microscopy (SEM, JEOL JSM 6060 and JEOL 7600F SEM with Oxford Inca EDS), atomic force microscopy (AFM, Nanosurf easyScan 2), x-ray diffraction (XRD, Rigaku 300 and Bruker D8 Discover), and transmission electron microscopy (TEM, JEOL JEM 1011).

RESULTS AND DISCUSSION

Initial experiments have been conducted using metal catalysts on Si and SiO_2/Si substrates. It is very important to study the changes experienced by the catalysts before the nanowire growth, since the nanowire growth occurs through the catalyst-assisted mechanism. Therefore, the morphology of the Ni film over the Si substrate was analyzed by AFM before (as-deposited) and after a heat treatment in H_2 and NH_3. It was observed that as-deposited Ni film on Si substrate has a relatively smooth surface. However, the surface morphology of the Ni film after the heat treatment at 1100°C in H_2 and NH_3 was extremely rough. The heat treatment results in formation of Ni islands and clusters, which promote growth of AlN nanowires. The size of these clusters can reach up to few microns, as measured by AFM.

Figure 1 shows SEM images of the AlN nanowires grown on 20-nm Ni coated Si and SiO_2/Si with 25 nm Co nanoparticles at 1100°C under H_2 as carrier gas. The AlN nanowire diameters grown on Ni-film are in the range of 20 nm to 30 nm and lengths about few microns. It is worth mentioning that the metal catalysts were observed at the end of the nanowires. This observation indicates that AlN nanowire growth has taken place via vapor-liquid-solid (VLS) growth mechanism. The AlN nanowire diameters grown with Co nanoparticles are in the range of 50 nm to 100 nm and lengths about several microns, as determined by the AFM measurements. Majority of the AlN nanowires grown with Co have kink morphology. The growth run has been repeated at the same conditions with different catalyst materials including Fe and Au. Nevertheless, Fe or Au catalysts did not yield any nanowire growth. Following that,

18

AlN growth was carried out at different temperatures with the most active catalysts (Ni and Co). In the temperature range of 800-950°C, no nanowire growth was observed. AlN nanowires appeared at 1000°C and the yield has increased with temperature and reaching its peak at 1100°C.

(a)   (b)

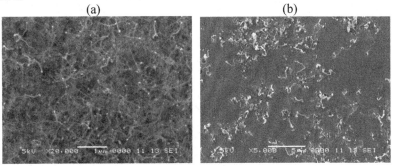

Figure 1. SEM images of AlN nanowires grown at 1100°C under H_2 with different catalysts: (a) Ni-film, (b) Co-nanoparticles (dilute solution).

Figure 2 shows SEM images of the catalyst-free grown ultra-dense AlN nanostructure films at 1100°C. The AlN nanostructures were directly grown on the source mixture of Al-Al_2O_3, which was placed directly to the alumina boat without any substrate. These free-standing nanostructure films were attached to conductive carbon tape before placing to the SEM sample holder. The AlN nanorod diameters are in the range of 60 nm to 90 nm and lengths up to few tens of microns. The mechanism for the catalyst-free growth of the AlN nanostructures is unclear. However, it is believed that the growth takes place through a VLS growth mechanism with Al droplets serving as the catalyst. Energy-dispersive X-ray spectroscopy (EDS) was used for the chemical analysis of the grown AlN nanostructures. The EDS spectrum, shown in Figure 3, indicates the presence of only Al and N elements. No oxygen was observed. This result indicates that the grown AlN nanostructures have high purity.

(a)   (b)

Figure 2. SEM images of catalyst-free grown ultra-dense AlN nanostructure films at 1100°C: (a) nanoneedles, (b) nanorods.

Next, XRD measurements were carried out using Cu K_α radiation ($\lambda = 1.54056$ A°) to determine the structure and crystal quality of the AlN nanostructures and a typical spectrum is shown in Figure 4. The diffraction peaks in the spectrum were indexed to a hexagonal wurtzite crystal structure. The lattice constants derived from the peak positions were a = 0.3114 nm and c = 0.4979 nm (JCPDS: No. 25-1133). The diffraction peaks and their positions from the AlN nanostructures can be listed as: (100) - 33.2°; (002) - 36.04°; (101) - 37.92°; (102) - 49.82°; (110) - 59.35°.

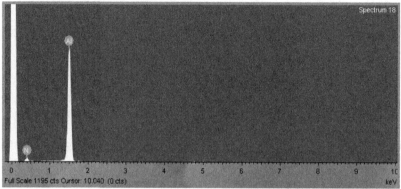

Figure 3. EDS spectra acquired from the catalyst-free grown AlN nanostructures.

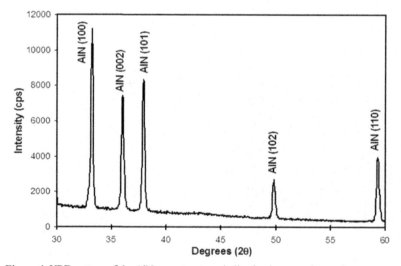

Figure 4. XRD pattern of the AlN nanostructures indicating hexagonal wurtzite structure.

No characteristic peak associated with other crystalline forms was detected in the XRD spectrum. These results suggest that the catalyst-free grown AlN nanostructures consist of only one crystalline phase. Furthermore, the narrow peaks indicate a narrow range of composition and stress; and the high intensity peaks indicate the presence of many crystalline AlN nanostructures (see figure 4).

Transmission electron microscopy was performed to characterize the morphology and diameter of the nanostructures. The TEM specimen was prepared by breaking small pieces from the dense film to the carbon-coated copper grids. Figure 5 shows the TEM image of the AlN nanorods with diameters ranging from 60 nm to 90 nm. Further, the diameter of some nanorods gradually increases along its length as observed by TEM.

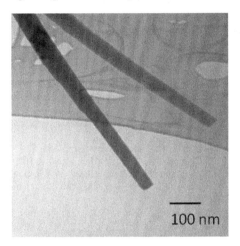

Figure 5. TEM image of the catalyst-free grown AlN nanorods.

CONCLUSIONS

We have successfully demonstrated both catalyst-assisted and catalyst-free growth of AlN nanostructures. The catalyst-assisted growth on Ni-coated film resulted in small diameter (20 – 30 nm) AlN nanowires with lengths of about few microns. In fact, the yield, as measured with the degree of surface coverage by nanowires, has been the highest for the Ni catalyst at 1100°C. Moreover, no significant growth has been observed at temperatures lower than 1000°C due to low vapor pressure of Al metal. The direct nitridation of mixture of Al-Al$_2$O$_3$ by NH$_3$ has resulted in very dense AlN nanostructures specifically nanoneedles and nanorods with diameters of 60 nm to 90 nm and lengths up to few tens of microns. Consequently, free-standing nanostructure films provide great opportunities due to enabling easy transfer of these into any substrate for the development of new devices for many technological applications including piezoelectric electricity generators.

ACKNOWLEDGMENTS

K.T. gratefully thanks to Appalachian Regional Commission (ARC) for providing financial support for this research. The authors are also grateful to Biology Department for providing the opportunity to do electron microscopy.

REFERENCES

1. S. Nakamura, *Science* **281**, 956 (1998).
2. S.G. Yang, S. Pakhomov, T. Hung, and C.Y. Wong, *Appl. Phys. Lett.* **81**, 2418 (2003).
3. S. Y. Wu, H.X. Liu, L. Gu, R.K. Singh, L. Budd, M. Van Schifgaarde, M.R. McCartney, D.J. Smith, and N. Newman, *Appl. Phys. Lett.* **82**, 3047 (2003).
4. V. Cimalla, Ch. Foerster, D. Cengher, K. Tonisch, and O. Ambacher, *Phys. Stat. Sol (b)* **243**, 1476 (2006).
5. V.N. Tondare, C. Ballasubramanian, S.V. Shende, D.S. Joag, V.P. Godbole, S.V. Bhoraskar, and M. Bhadbhade, *Appl. Phys. Lett.* **80**, 4813 (2002).
6. J.A. Haber, O.C. Gibbons, and W.E. Buhro, *Chem. Mater.* **10**, 4062 (1998).
7. Y.J. Zhang, J. Liu, R.R. He, Q. Zhang, X.Z. Zhang, and J. Zhu, *Chem. Mater.* **13**, 3899 (2001).
8. T. Suehiro, J. Tatami, T. Meguro, K. Komeya, and S. Matsuo, *J. Am. Ceram. Soc.* **75**, 910 (2002).
9. H.M. Wu, J.Y. Liang, K.L. Lin, and C.C. Chou, *Ferroelectrics* **383**, 73 (2009).
10. Q. Wu, Z. Hu, X.Z. Wang, Y. Chen, and Y. N. Lu, *J. Phys. Chem. B* **107**, 9726 (2003).
11. Y.B. Tang, H.T. Cong, Z.M. Wang, and H.M. Cheng, *Chem. Phys. Lett.* **416**, 171 (2005).
12. C. Liu, Z. Hu, Q. Wu, X.Z. Wang, Y. Chen, H. Sang, J.M. Zhu, S.Z. Deng, and N.S. Xu, *J. Am. Ceram. Soc.* **127**, 1318 (2005).
13. Q. Zhao, H.Z. Zhang, X.Y. Xu, Z. Wang, J. Xu, D.P. Yu, G.H. Li, and F.H. Su, *Appl. Phys. Lett.* **86**, 193101 (2005).
14. S.C. Shi, C.F. Chen, S. Chattopadhyay, Z.H. Lan, K.H. Chen, and L.C. Chen, *Adv. Funct. Mater.* **15**, 781 (2005).
15. F. Liu, Z.J. Su, F.Y. Mo, L. Li, Z.S. Chen, Q.R. Liu, J. Chen, S.Z. Deng, and N.S. Xu, *Nanoscale* **3**, 610 (2011).
16. 12. X. Wang, J. Song, F. Zhang, C. He, Z. Hu, and Z. Wang, *Adv. Mater.* **22**, 1 (2010).

Mater. Res. Soc. Symp. Proc. Vol. 1324 © 2011 Materials Research Society
DOI: 10.1557/opl.2011.840

Enhanced Light Emission at Self-assembled GaN Inversion Domain Boundary

Mei-Chun Liu[1], Yuh-Jen Cheng,[1] Jet-Rung Chang,[2] and Chun-Yen Chang[2]
[1]Research Center for Applied Sciences, Academia Sinica, Taipei 11529, Taiwan
[2]Institute of Electronics, National Chiao Tung University, 1001 Ta Hsueh Rd., Hsinchu 300, Taiwan

ABSTRACT

We report the fabrication of GaN lateral polarity inversion heterostructure with self assembled crystalline inversion domain boundaries (IDbs). The sample was fabricated by two step molecular-beam epitaxy (MBE) with microlithography patterning in between to define IDbs. Despite the use of circular pattern, hexagonal crystalline IDbs were self assembled from the circular pattern during the second MBE growth. Both cathodoluminescent (CL) and photoluminescent (PL) measurements show a significant enhanced emission at IDbs and in particular at hexagonal corners. The ability to fabricate self assembled crystalline IDbs and its enhanced emission property can be useful in optoelectronic applications.

INTRODUCTION

III-Nitride semiconductors, in particular GaN, have attracted great research interests in past few years due to their promising applications for UV to blue optoelectronic devices [1]. One of the important properties of wurtzite GaN is its strong spontaneous and piezoelectric polarization, which can induce surface charges and create large internal electric field in the film. This electric field can strongly affect the electrical and optical properties of GaN based devices [2, 3]. It can reduce the electron-hole wave function overlap in quantum well devices and have adverse effect on its light emitting efficiency. The direction of internal field depends on crystal c-axis orientation which is not symmetric. The c-axis crystal surface is conventionally labeled as Ga- (0001) or N-polar (000-1).

Recently, there are interests in studying the physical property of inversion domain boundary between Ga- and N-polar regions. The IDB has been found to exist at microscopic scale in Ga-polar GaN thin film grown by MBE [4]. The IDB can also be created by intentionally growing lateral polarity heterostructures, where patterned Ga- and N-polar regions are laterally grown on the same substrate to form IDbs [5]. The ability to fabricate controlled polarity pattern can open up new dimension for GaN device design and applications. The IDB has been studied theoretically [6, 7], and experimentally imaged by high-resolution transmission electron microscope [4, 5], and piezoelectric force microscope [8, 9]. It was theoretically predicted that IDbs in wurtzite GaN would not have electronic states in the band gap, implying that they would not affect PL efficiency [6]. The optical property of IDB has been investigated by high resolution spatially resolved PL measurement [10]. The measurement was done at 10K low temperature and significantly brighter emission was reported at IDbs. However, the observed emission from IDB was neither spectrally nor spatially uniform. The non-uniformity could be due to the coexistence of different crystalline planes in the intentionally patterned IDB, which was suggested possibly having mixed crystalline planes {10-10} and {11-20}. The fabricated IDB was a straight line in tens of μm and not oriented along any specific crystal plane. That could be the reason for leading to mixed crystalline planes in IDBs. The observed PL peaks have some variations and are about 30-40 meV lower than the bulk Ga-polar emission. A theoretical calculation however predicts a zero shift in emission peak at IDB [7]. The discrepancy may come from the mixed crystal planes in IDBs and possible defects associated with them. The ability to fabricate crystalline IDB can better reveal its physical property. It may also open up new applications due to its enhanced emission property.

Here we report the fabrication of self assembled IDB along {10-10} crystalline plane and the observation of enhanced light emission at IDB by CL and PL measurement. The crystalline IDBs were fabricated by two step rf-plasma-enhanced MBE, where microlithography patterning process was used in between two growth steps to define lateral polarity heterostructure boundary. Circular patterns were used in microlithography patterning process. However, hexagonal crystalline IDBs along {10-10} crystalline planes were self assembled from the original circular pattern after the second MBE re-growth. Both CL and PL measurements show an enhanced light emission at IDBs, in particular at hexagonal corners, and a zero shift in emission peak.

EXPERIMENT

It was well known that GaN grown on c-plane sapphire by MBE epitaxy normally has an N-polar surface. It was also demonstrated that the polarity of GaN can be switched to Ga-polar surface by pre-growing a high temperature AlN buffer layer on c-plane sapphire surface. The polarities of these two growths have been shown to correlate to in situ reflection high-energy electron diffraction (RHEED) patterns, where N- and Ga-polar surface shows (3x3) and (2x2) RHEED pattern, respectively [11-13]. The fabrication steps use these two growth techniques and microlithography patterning to grow patterned lateral polarity heterostructure, as shown in Fig. 1 (a)-(d). First, a high temperature (930 °C) thin AlN buffer layer (30 nm) was grown on top of a (0001) sapphire substrate followed by a thin layer of GaN growth (810 °C, 50 nm) (Fig. 1 (a)). Ga-polar GaN was grown on AlN buffer layer, which was confirmed by in situ (2x2) RHEED pattern. Photoresist (PR) was then spun on the sample and patterned using microlithography to create 3 μm circular mesa masks (Fig. 1 (b)). The sample was etched in an inductively coupled plasma (ICP) etcher using Ar/Cl2 gas until the unmasked GaN regions were etched down to expose the sapphire substrate. The PR mask was then removed, leaving 3 μm circular Ga-polar GaN mesas on sapphire substrate (Fig. 1(c)). The sample was subsequently reintroduced into the MBE system to grow 1 μm (810 oC) GaN layer. Since AlN buffer layer was removed by the etching process, the re-grown GaN on the exposed sapphire surface would be N-polar [11]. The same second step growth parameter was applied to a separate blank sapphire substrate to verify that it did grow N-polar GaN, which was confirmed by in situ (3x3) RHEED pattern. The GaN grown on the circular Ga-polar GaN mesa, which worked as a seeding layer, will still follow the same Ga polarity direction. This is true because it is as if the growth of Ga-polar GaN is interrupted and put back into regrowth under same growth condition, which would still follow Ga-polar growth direction. As a result, we obtained a patterned lateral polarity heterostructure with pre-defined IDBs.

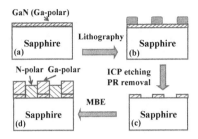

Fig. 1 (a)-(d) Processing flow for fabricating patterned polarity IDB. (a) Grow Ga-polar GaN layer. (b) Photoresist patterning. (c) ICP etching to remove unmasked region followed by PR removal (d). The second growth to form polarity heterostructure.

A scanning electron microscope (SEM) plane view of the sample is shown in Fig. 2 (a). It is interesting to observe the self assembly of hexagonal shapes after the second MBE growth even though originally circular pattern was used to define polarity inversion boundary. From the growth process design, the polarity inside hexagon is Ga-polar, while the surrounding area is N-polar. The IDBs were self assembled at {10-10} hexagonal crystal planes. The SEM was switched to cathodoluminescent detection under the same magnification condition. The CL image is shown in Fig. 2 (b). The focused e-beam spot size is less than 10 nm, which defines the CL imaging spatial resolution. The cathodoluminescent intensity is significantly stronger at IDBs and in particular strongest at hexagonal corners. The luminescent intensity at IDB gradually decreases with position moving away from corner along the boundary. An atomic force microscope scan shows that the hexagonal Ga-polar region is about 80 nm higher than the surrounding N-polar surface, which is consistent with the growth process design. The scanned height profile shows a slow inclined side wall of about 160 nm in lateral direction. To clarify if the stronger emission at IDB was related to the height difference between the two opposite polarity regions, similar fabrication steps but without AlN buffer layer to create polarity inversion were carried out to fabricate a GaN substrate with similar surface topography. First, a thin layer of GaN (810 °C, 80 nm) was grown on sapphire substrate. A similar photolithography process was used to pattern 3 μm circular PR mesas on the substrate. The photoresist mesas were used as etching masks in the subsequent ICP etching process to etch the unmasked region down to sapphire and leave 3 μm circular GaN mesas on sapphire substrates. The sample was then reintroduced into MBE to grow 1 μm (810 °C) GaN layer. A GaN substrate with similar topography as the previous substrate but without polarity inversion boundaries was fabricated. The CL image of such a sample did not show bright emission at the patterned boundaries. It thus confirmed that the strong emission observed in the lateral polarity heterostructure sample is related to polarity inversion boundary.

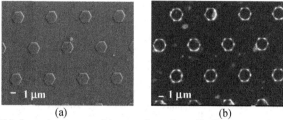

(a)   (b)

Fig. 2 (a) SEM plane view image of the sample surface. (b) CL image at the same magnification showing the enhanced luminecent intensity at IDBs.

The CL intensity spectra taken at the N-polar region, center of hexagonal Ga-polar region, middle of hexagonal boundary, and hexagonal corner are show in Fig. 3 (a). The full width at half maximum (FWHM) linewidth is about 14 nm for all spectra. The intensity at Ga-polar region is significantly higher than that at N-polar region. Ga-polar GaN is known to have better material quality than N-polar GaN, which explains the observed luminescent intensity difference between Ga- and N-polar region. The intensity at IDB, in particular at the hexagonal corner, is much higher than both Ga- and N-polar region. The peak wavelength of the enhanced luminescence at IDB is the same as those for Ga- and N-polar GaN, consistent with the theoretical prediction [10]. This is in contrast to the 30 meV redshift reported from IDB with mixed crystal planes where the redshift could be due to shallow defect traps created among the mixed crystal planes. The PL spectra of the sample were also measured using 325 nm He-Cd laser pumping source. The laser was focused on the sample by a 39X UV objective with a focused spot size of 0.8 μm. The PL spectra at similar locations are shown in Fig. 3 (b). It reconfirms the enhanced light emission at IDBs. The order of intensity magnitudes at different locations is similar to the CL case except that the intensity at corner is not as strong. This is probably due to the much larger excitation source spot size in PL measurement and implies that the great enhancement at corner is localized to a very small area. A theoretical model based on ab initio density functional calculation has been proposed to explain the enhancement [7], which shows that the calculated electronic potential tends to trap electrons and holes in the IDB vicinity, therefore increases the recombination efficiency. A heuristic model of polarization field cancelation at IDB due to opposite polarity across the boundary has also been proposed to qualitatively explain the electron-hole recombination efficiency enhancement.

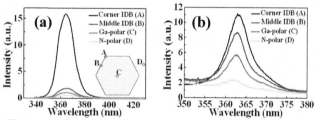

Fig. 3 (a) CL intensity spectra at various sample locations. (b) PL intensity spectra at various sample locations.

CONCLUSIONS

In summary, the fabrication of polarity inversion GaN heterostructure with self assembled {10-10} crystal plane IDBs is demonstrated. The fabrication process involves two step rf-enhanced MBE growths and lithography patterning in between two growths to define IDBs. Despite the use of circular patterns for defining IDBs, hexagonal crystalline IDBs are self assembled from the original circular patterns in the second epitaxial growth. CL and PL measurements show that luminescent property is significantly enhanced at IDBs. The ability to fabricate crystalline IDB and the demonstration of its enhanced luminescent property offers new device design dimension and may open up new optoelectronic applications.

ACKNOWLEDGMENTS

This work was financially supported by Sinica Nano-program and in part by the National Science Council of Republic of China (ROC) Taiwan under contract NSC97-2112-M-001-027-MY3 and Sinica Nano Program.

REFERENCES

1. T. Mukai, K. Takekawa, and S. Nakamura, Jpn. J. Appl. Phys., Part 2 37, L839 (1998).
2. O. Ambacher, J. Smart, J. R. Shealy, N. G. Weimann, K. Chu, M. Murphy, R. Dimitrov, L. Wittmer, M. Stutzmann, W. Rieger, and J. Hilsenbeck, J. Appl. Phys. 85, 3222 (1999).
3. M. Stutzmann, O. Ambacher, M. Eickhoff, U. Karrer, A. Lima Pimenta, R. Neuberger, J. Schalwig, R. Dimitrov, P. Schuck, and R. Grober, Phys. Status Solidi B 288, 505 (2001).
4. C. Iwamoto, X. Q. Shen, H. Okumura, H. Matuhata, and Y. Ikuhara, Appl. Phys. Lett. 79, 3941 (2001).
5. F. Liu, R. Collazo, S. Mita, Z. Sitar, G. Duscher, and S. J. Pennycook, Appl. Phys. Lett. 91, 203115 (2007).
6. J. E. Northrup, J. Neugebauer, and L. T. Romano, Phys. Rev. Lett. 77, 103 (1996).
7. V. Fiorentini, Appl. Phys. Lett. 82, 1182 (2003).
8. B. J. Rodriguez, A. Gruverman, A. I. Kingon, R. J. Nemanich, and O. Ambacher, Appl. Phys. Lett. 80, 4166 (2002).
9. Ryuji Katayama,a Yoshihiro Kuge, Kentaro Onabe, Tomonori Matsushita and Takashi Kondo, Appl. Phys. Lett. 89, 231910 (2006).
10. P. J. Schuck, M. D. Mason, R. D. Grober, O. Ambacher, A. P. Lima, C. Miskys, R. Dimitrov, and M. Stutzmann, Appl. Phys. Lett. 79, 952 (2001).
11. X. Q. Shen, T. Ide, S. H. Cho, M. Shimizu, S. Hara, and H. Okumura, Appl. Phys. Lett. 77, 4013 (2000).
12. D. Huang, P. Visconti, K. M. Jones, M. A. Reshchikov, F. Yun, A. A. Baski, T. King, and H. Morkoc, Appl. Phys. Lett. 78, 4145 (2001).
13. A. R. Smith, R. M. Feenstra, D. W. Greve, M.-S. Shin and M. Skowronski, J. Neugebauer, and J. E. Northrup, Appl. Phys. Lett. 72, 2114 (2001).

Mater. Res. Soc. Symp. Proc. Vol. 1324 © 2011 Materials Research Society
DOI: 10.1557/opl.2011.1056

AlGaN Channel HEMT with Extremely High Breakdown Voltage

Takuma Nanjo[1], Misaichi Takeuchi[2], Akifumi Imai[1], Yousuke Suzuki[1], Muneyoshi Suita[1], Katsuomi Shiozawa[1], Yuji Abe[1], Eiji Yagyu[1], Kiichi Yoshiara[1] and Yoshinobu Aoyagi[2]
[1]Mitsubishi Electric Corporation, Advanced Technology R & D Center, 8-1-1, Tsukaguchi-honmachi, Amagasaki, Hyogo 661-8661, JAPAN
[2]Ritsumeikan Global Innovation Research Organization, Ritsumeikan University, 1-1-1 Noji-Higashi, Kusatsu, Shiga 525-8577, JAPAN

ABSTRACT

A channel layer substitution of a wider bandgap AlGaN for a conventional GaN in high electron mobility transistors (HEMTs) is an effective method of enhancing the breakdown voltage. Wider bandgap AlGaN, however, should also increase the ohmic contact resistance. Si ion implantation doping technique was utilized to achieve sufficiently low resistive source/drain contacts. The fabricated AlGaN channel HEMTs with the field plate structure demonstrated good pinch-off operation with sufficiently high drain current density of 0.5 A/mm without noticeable current collapse. The obtained maximum breakdown voltages was 1700 V in the AlGaN channel HEMT with the gate-drain distance of 10 μm. These remarkable results indicate that AlGaN channel HEMTs could become future strong candidates for not only high-frequency devices such as low noise amplifiers but also high-power devices such as switching applications.

INTRODUCTION

GaN-based wide bandgap semiconductors have superior electrical properties such as a high electron saturation velocity and a high breakdown field. In particular AlGaN/GaN high-electron mobility transistors (HEMTs) with high-density two-dimensional electron gas have been widely studied for high-power and high-frequency applications.[1-10] To increase a power density of HEMTs, it is a simple strategy to increase a breakdown voltage. Therefore applying a wider bandgap material for a channel layer, which generally enhances the breakdown field (E_c), is one of effective methods. Higher Al composition AlGaN is an available material to increase the breakdown voltage without decreasing a drain current density, because the E_c of AlN, which has about twice wide of the energy bandgap comparing to that of GaN, is about four times larger than that of GaN, and the electron saturation velocity of AlN is almost as same as that of GaN as shown in table I. Although the AlGaN channel HEMTs have high potential to increase the breakdown voltage, the wider bandgap leads to the difficulty to form a sufficiently low resistive ohmic contact. Therefore a breakthrough technology to reduce the contact resistance is required. In order to solve this problem, we employed Si ion implantation doping technique that we have developed for conventional GaN channel HEMTs.[11]

In this paper, we report that AlGaN channel HEMTs, where we utilized Si ion implantation doping technique to achieve sufficiently low resistive contacts, were operated with remarkably enhanced breakdown voltage and sufficiently high drain current density without noticeable current collapse.

Table I. Properties of semiconductor materials using HEMTs

	GaAs	GaN	AlN
Bandgap, E_g (eV)	1.42	3.4	6.2
Electric Breakdown Field, E_c (V/cm)	0.4×10^6	3.3×10^6	12×10^6
Electron Mobility, μ (cm^2/Vs)	8500	2000	1090
Electron Saturation Velocity, V_{sat} (cm/s)	2.0×10^7	2.5×10^7	2.2×10^7
Johnson's figure of merit	7	760	7800
Baliga's figure of merit	9	39	67

EXPERIMENT

Figure 1 shows a cross-sectional structure of fabricated $Al_yGa_{1-y}N/Al_xGa_{1-x}N$ HEMTs. Unintentionally doped epitaxial layers were grown by metalorganic chemical vapor deposition. We prepared three types of $Al_yGa_{1-y}N/Al_xGa_{1-x}N$ heteroepitaxial wafers. As shown in Table II, these were one GaN channel structure on Sapphire substrate and two AlGaN channel structures on Sapphire and SiC substrates, respectively. A barrier layer thickness and a difference of Al composition between the barrier and the channel layer were equivalent in three wafers.

Figure 2 shows a fabrication process of the $Al_yGa_{1-y}N/Al_xGa_{1-x}N$ HEMTs. The fabrication process started with selective Si ions implantation into source/drain regions. ^{28}Si ions were implanted with an energy of 50 keV at a dose concentration of 1 x 10^{15} cm^{-2} at room temperature, and subsequently implanted Si ions were activated by rapid thermal annealing at 1200 ^{O}C for 5 min in an environment of flowing nitrogen. During the Si ions implantation and the activation annealing, wafer surfaces were capped by 30 nm thick of SiN layer deposited by plasma-enhanced chemical vapor deposition. After removing the SiN cap layer in a HF solution, source/drain ohmic contacts were formed by Ti/Al metallization and subsequent rapid thermal annealing at 600 ^{O}C for 2 min in an environment of flowing nitrogen. Then device isolation was performed by Zn ion implantation[12] and Ni/Au Schottky gate contact was formed. The fabrication of HEMTs was completed by depositing a SiN dielectric film using catalytic chemical vapor deposition[13-15] and subsequent formation of a Ni/Au field plate which was connected to the gate contact at an electrode pad. All electrodes were defined by a conventional photolithography and liftoff technique, and the all metals were deposited by an electron beam evaporation system. In fabricated $Al_yGa_{1-y}N/Al_xGa_{1-x}N$ HEMTs, the gate length (L_g), the gate width (W_g), the distance between the source and gate (L_{sg}), and the distance between the gate and drain (L_{gd}) were 1, 50, 1, and 2~10 μm, respectively.

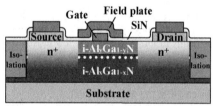

Figure 1. Schematically illustrated cross-sectional view of fabricated $Al_yGa_{1-y}N/Al_xGa_{1-x}N$ HEMTs.

Table II. RMS values, sheet carrier concentrations, electron mobilities, and sheet resistances for three $Al_yGa_{1-y}N/Al_xGa_{1-x}N$ heterostructures.

Barrier layer	$Al_{0.25}Ga_{0.75}N$	$Al_{0.40}Ga_{0.60}N$	$Al_{0.40}Ga_{0.60}N$
Barrier layer thickness (nm)	25	25	25
Channel layer	GaN	$Al_{0.15}Ga_{0.85}N$	$Al_{0.15}Ga_{0.85}N$
Substrate	Sapphire	Sapphire	4H-SiC
RMS (nm)	0.26	0.68	0.41
Sheet carrier concentration (cm^{-2})	8.7×10^{12}	7.9×10^{12}	9.4×10^{12}
Electron mobility (cm^2/Vs)	1550	460	510
Sheet resistance (Ω/sq)	460	1720	1310

- **Epitaxial layer growth**
  i-$Al_yGa_{1-y}N$/ i-$Al_xGa_{1-x}N$ hetero structure
- **Source/Drain Ohmic contact**
  Si ion implantation
  Activation annealing
  Ohmic metal (Ti/Al) deposition
  Alloying annealing
- **Device Isolation**
  Zn ion implantation
- **Gate Schottky contact**
  Schottky metal (Ni/Au) deposition
- **Field plate**
  Dielectric film (SiN) deposition
  Field plate metal (Ni/Au) deposition

Figure 2. Fabrication process of $Al_yGa_{1-y}N/Al_xGa_{1-x}N$ HEMTs.

RESULTS AND DISCUSSION
Eptaxial layer characteristics

Sheet carrier concentrations, electron mobilities and sheet resistances of the three wafers were evaluated from a Hall measurements at room temperature as summarized in Table II. The sheet carrier concentration was decreased by increasing the Al composition in epitaxial layers and increased by substituting the substrate to SiC, which were attributed to crystal quality degradation by increasing the Al composition and improvement by substituting the substrate to SiC mainly due to less lattice mismatch with the substrate. As proof of the alteration in crystal quality, a surface morphology and a RMS value from AFM images were degraded by increasing the Al composition and improved by substitution of the substrate to SiC as shown Fig. 2 and table II. Electron mobility was also decreased by increasing the Al composition in epitaxial layers which was attributed to an increase of alloy scattering and this degradation is one of disadvantages in AlGaN channel HEMTs in presence. The electron mobility in AlGaN channel structure, however, was higher than the calculated value[16] and it was improved by substituting the substrate to SiC due to the improvement in crystal quality. The obtained sheet resistances were sufficiently low to operate the HEMTs.

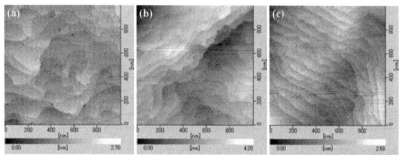

Figure 3. AFM images of (a) $Al_{0.25}Ga_{0.75}N/GaN$ layers on Sapphire substrate, (b) $Al_{0.40}Ga_{0.60}N/Al_{0.15}Ga_{0.85}N$ layers on Sapphire substrate and (c) $Al_{0.40}Ga_{0.60}N/Al_{0.15}Ga_{0.85}N$ layers on SiC substrate.

Ohmic characteristics

Figure 4 shows current-voltage (I-V) curves between two ohmic electrodes spaced 4μm with and without Si ion implantation which were formed on the $Al_{0.40}Ga_{0.60}N/Al_{0.15}Ga_{0.85}N$ epitaxial layers on SiC substrate. In the sample without Si ion implantation, a contact resistance was so high that we could not calculate it from a circular transfer length method (CTLM), because the AlGaN barrier layer was not only unintentionally doped but also the Al composition was higher comparing to the conventional barrier layer of AlGaN/GaN structure. On the other hand, in the sample with Si ion implantation, we could realize a low ohmic contact resistance of 1.5×10^{-5} Ωcm^2 led from the CTLM. As a result Si ion implantation technique was very effective to obtain sufficiently low contact resistance to operate the transistor with this AlGaN/AlGaN hetero-structure.

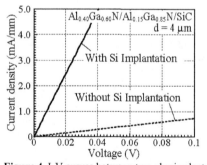

Figure 4. I-V curves between two ohmic electrodes with and without Si ion implantation.

32

HEMT characteristics

Figure 5 shows a drain current-drain voltage (I_d-V_d) curves measured in dc mode (dashed lines) and pulsed mode (solid lines) in fabricated three HEMTs. The L_g, W_g, L_{sg}, and L_{gd} of the measured HEMTs were 1, 100, 1, and 2 μm, respectively. In all of the HEMTs, we realized transistor operation with good pinch-off characteristics. Although drain current density decreased by the increase of Al composition in epitaxial layers due to the deterioration of electron mobility, maximum values obtained at the gate voltage (V_g) of +2 V were sufficiently high of 0.34 and 0.50 A/mm in the AlGaN channel HEMTs on Sapphire and SiC substrate, respectively. The degradations in the operation of pulsed mode, that is current collapse, were similar in both hetero-structures on the Sapphire substrate. In the AlGaN channel HEMTs on the SiC substrate, however, relatively large current collapse occurred. We consider this current collapse was attributed to residual carriers in the deep region of the channel layer whose possibility will be discussed below.

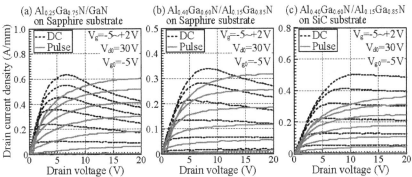

Figure 5. DC and Pulse I_d-V_d curves of (a) $Al_{0.25}Ga_{0.75}N/GaN$ HEMT on Sapphire substrate, (b) $Al_{0.40}Ga_{0.60}N/Al_{0.15}Ga_{0.85}N$ HEMT on Sapphire substrate and (c) $Al_{0.40}Ga_{0.60}N/Al_{0.15}Ga_{0.85}N$ HEMT on SiC substrate. The L_g, W_g, L_{sg}, and L_{gd} of the measured HEMTs were 1, 100, 1, and 2 μm, respectively.

By increasing the Al composition in hetero structure, drain breakdown voltage was drastically enhanced. Figure 6 shows V_d dependences of I_d and gate current (I_g) at off-state (V_g = -10 V) in the fabricated three HEMTs with the L_{gd} of 3 and 10 μm, respectively. In the HEMTs with the L_{gd} of 3 μm, which are commonly used for high-frequency devices such as low noise amplifiers, obtained breakdown voltage in AlGaN channel HEMT on Sapphire substrate was 500 V which was higher than that in GaN channel HEMT. In the HEMTs with the L_{gd} of 10 μm, which are commonly used for high-power devices such as switching applications, we realized an extremely high breakdown voltage of 1.7 kV in AlGaN channel HEMT on Sapphire substrate which was much higher than that in GaN channel HEMT. To our knowledge, these are the highest breakdown voltages for group III nitride based HEMTs with similar device dimensions.[2-10] The higher breakdown field due to the wider bandgap contributed to these remarkable breakdown voltage enhancements in AlGaN channel HEMTs. By substituting the substrate to SiC, breakdown voltages decreased regardless of L_{gd} due to the relatively large drain leakage current in high V_d. Similar drain leakage current had occurred in the AlGaN channel HEMTs using GaN buffer layer on Sapphire substrate and detailed investigation had revealed an existence of residual carriers in the deep region of the channel layer that formed a leakage current pass from source to drain at off-state.[17] We consider, in this study, similar residual carriers existed in the deep region of the channel layer due to the substitution of substrate from Sapphire to SiC, and these residual carriers would also trap electrons in the pulsed mode operation caused current collapse. Although breakdown voltages were decreased, the obtained values were 380 and 1080 V in the HEMTs with the L_{gd} of 3 and 10 μm, respectively, and these were sufficiently higher than that in GaN channel HEMTs.

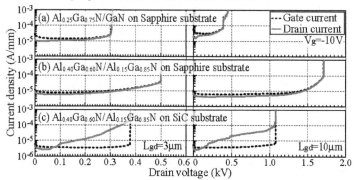

Figure 6. Off-state (V_g = -10 V) I_d-V_d and I_g-V_d curves of (a) $Al_{0.25}Ga_{0.75}N$/GaN HEMT on Sapphire substrate, (b) $Al_{0.40}Ga_{0.60}N$/$Al_{0.15}Ga_{0.85}N$ HEMT on Sapphire substrate and (c) $Al_{0.40}Ga_{0.60}N$/$Al_{0.15}Ga_{0.85}N$ HEMT on SiC substrate with the L_{gd} of 3 and 10 μm. The L_g, W_g, and L_{sg} of the measured HEMTs were 1, 100, and 1 μm, respectively.

Finally we compared the obtained specific on-state resistances and breakdown voltages to the state-of-the-art values[3-10] as shown in Fig. 7. These values were obtained in the fabricated three HEMTs with different L_{gd}. We demonstrated that presented AlGaN channel HEMTs were sufficiently competitive to the GaN based devices. We believe that farther development in AlGaN channel HEMTs should realize the high performances exceeding the GaN based devices.

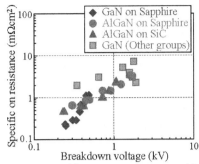

Figure 7. Specific on-state resistances and breakdown voltages in fabricated HEMTs comparing with those in state-of-the-art GaN-based devices.

CONCLUSIONS

Remarkable breakdown voltage enhancements were demonstrated in the AlGaN channel HEMTs. A Si ion implantation doping technique was utilized to reduce an increased ohmic contact resistivity due to applying a high Al mole fractional AlGaN barrier layer. The fabricated AlGaN channel HEMTs with field plate structure demonstrated good pinch-off operation and sufficiently high drain current density over 0.5 A/mm without noticeable current collapse. The obtained breakdown voltage was 1700 V in the AlGaN channel HEMTs with the L_{gd} of 10 μm. These remarkable results indicate that AlGaN channel HEMTs could become future strong candidates for not only high-frequency devices such as low noise amplifiers but also high-power devices such as switching applications.

ACKNOWLEDGMENTS

The authors would like to thank Mr. H. Koyama, Mr. Y. Kamo, and Mr. Y. Nakao for their collaboration in this study.

REFERENCES

1. S. Keller, Y-F. Wu, G. Parish, N. Ziang, J. J. Xu, B. P. Keller, S. P. DenBaars and U. K. Mishra, IEEE Trans. Electron Devices **48**, 552 (2001).
2. T. Kikkawa, Jpn. J. Appl. Phys., Part 1 **44**, 4896 (2005).
3. M. Hikita, M. Yanagihara, K. Nakazawa, H. Ueno, Y. Hirose, T. Ueda, Y. Uemoto, T. Tanaka, D. Ueda and T. Egawa, IEEE Trans. Electron Devices **52**, 1963 (2005)
4. Y. Dora, A. Chakraborty, L. McCarthy, S. Keller, S. P. DenBaars and U. K. Mishra, IEEE Electron Device Lett. **27**, 713 (2006)

5. N. Tipirneni. A. Koudymov, V. Adivarahan, J. Yang, G. Simin and M. A. Khan, IEEE Electron Device Lett. **27**, 716 (2006)
6. C. S. Suh, Y. Dora, N. Fichtenbaum, L. McCarthy, S. Keller, and U. K. Mishra, Tech. Dig. - Int. Electron Devices Meet. (2006).
7. Y. C. Choi, J. Shi, M. Pophristic, M. G. Spencer and L. F. Eastman, J. Vac. Sci. Technol. B **25**, 1836 (2007)
8. Y. Uemoto, D. Shibata, M. Yanagihara, H. Ishida, H. Matsuo, S. Nagai, N. Batta, M. Li, T. Ueda, T. Tanaka, and D. Ueda, Tech. Dig. - Int. Electron Devices Meet. (2007).
9. T. Morita, M. Yanagihara, H. Ishida, M. Hikita, K. Kaibara, H. Matsuo, Y. Uemoto, T. Ueda, T. Tanaka and D. Ueda, Tech. Dig. - Int. Electron Devices Meet. (2007).
10. N. Ikeda, Y. Niiyama. H. Kambayashi, Y. Sato, T. Nomura, S. Kato and S. Yoshida, Proc. IEEE, **98**, 1151 (2010)
11. M. Suita, T. Nanjo, T. Oishi, Y. Abe and Y. Tokuda, Phys. Status Solidi C **3**, 2364 (2006)
12. T. Oishi, N. Miura, M. Suita, T. Nanjo, Y. Abe, and T. Ozeki, H. Ishikawa, T. Egawa and T. Jimbo, J. Appl. Phys. **94**, 1662 (2003).
13. Y. Kamo, T. Kunii H. Takeuchi, Y. Yamamoto, M. Totsuka, T. Shiga, H. Minami, T. Kitano, S. Miyakuni, T. Oku, A. Inoue, T. Nanjo, Y. Tsuyama, R. Shirahana, K. Iyomasa, K. Yamanaka, T. Ishikawa, T. Takagi, K. Marumoto and Y. Matsuda, 2005 IEEE MTT-S Int. Microwave Symp. Dig. **495** (2005).
14. M. Higashiwaki, N. Hirose and T. Matsui: IEEE Electron Device Lett. **26, 139** (2005).
15. M. Higashiwaki, T. Matsui and T. Mimura: IEEE Electron Device Lett. **27**, 16 (2006).
16. A. Raman, S. Dasgupta, S. Rajan, J. S. Speck and U. K. Mishra, Jpn. J. Appl. Phys., **47**, 3357 (2008)
17. T. Nanjo, M. Takeuchi, A. Imai, M. Suita, T. Oishi, Y. Abe, E. Yagyu, T. Kurata, Y. Tokuda and Y. Aoyagi, Electron. Lett. **39**, 750 (2003).

Mater. Res. Soc. Symp. Proc. Vol. 1324 © 2011 Materials Research Society
DOI: 10.1557/opl.2011.962

Fabrication and Optical Properties of Green emission semipolar {10$\bar{1}$1} InGaN/GaN MQWs Selective Grown on GaN Nanopyramid Arrays

Shih-Pang Chang[1,2], Jet-Rung Chang[3], Ji-Kai Huang[1], Jinchai Li[1,4, a)], Yi-Chen Chen[1], Kuok-Pan Sou[1], Yun-Jing Li[5], Hung-Chih Yang[2], Ta-Cheng Hsu[2],Tien-Chang Lu[1], Hao-Chung Kuo[1, b)] and Chun-Yen Chang[3]

[1]Department of Photonics & Institute of Electro-Optical Engineering, National Chiao Tung University, 1001 Ta Hsueh Rd., Hsinchu 300, Taiwan

[2]R&D Division, Epistar Co. Ltd., Science-based Industrial Park, Hsinchu 300, Taiwan

[3]Department of Electronic Engineering, National Chiao Tung University, 1001 Ta Hsueh Rd., Hsinchu 300, Taiwan

[4]Department of Physics, Xiamen University, Xiamen 361005, China

[5]Institute of Lighting and Energy Photonics, National Chiao Tung University at Tainan, Taiwan

Abstract

We report that the high crystalline and high efficiency green emission semipolar {10$\bar{1}$1} InGaN/GaN multiple quantum wells (MQWs) grown on the {10$\bar{1}$1} facets of GaN nanopyramid arrays by selective area epitaxy. Clear and sharp interfaces of the semipolar {10$\bar{1}$1} InGaN/GaN MQWs was observed by transmission electron microscopy images. As comparing with (0001) MQWs, the internal electric field of {10-11} MQWs was remarkably reduced from 1.7 MV/cm to 0.5 MV/cm, and the room temperature (RT) internal quantum efficiency (IQE) at green emission was enhanced by about 80%. This greatly enhancement of IQE is due to suppress the polarization effect in the {10$\bar{1}$1} MQWs which shorten the radiative recombination to compete with nonradiative recombination at RT. These results evince that the {10$\bar{1}$1} planes are promising for solving the efficiency green gap of III-nitride light emitters.

Introduction

The III-nitride materials have attracted much attention due to its tremendous potential for fabricating light emitting diodes (LEDs) with a full color application from UV (~3.4 eV) to near infrared (~0.7 eV) [1]. Researchers have realized the high efficiency blue LEDs with conventional c-plane InGaN/GaN quantum wells (QWs) in recent years. However, the efficiency of green LEDs is still very low and far from adequate for application. This is the so-called efficiency green gap. As has been reported, the so-called efficiency green gap can be mainly attributed to the strain induced polarization field in the InGaN/GaN QWs which results in the quantum confine stark effect (QCSE) and therefore the reduction of recombination rate of carriers [2]. To date, tremendous efforts have been made to solve QCSE obstacle by using the nonpolar InGaN/GaN structure grown on various substrates [3].

However, the heteroepitaxy of nonpolar GaN still suffers from highly threading dislocation density (TDD), and the ideal solutions for the green efficiency gap is still lacking [2-3]. GaN based nanorod LEDs is an attractive approach towards achieving high efficiency LEDs [4-5], which possess lower extended defect, accommodate the large lattice mismatch between sapphire substrate and GaN epilayer [6], and highly light extraction structure properties [4-5]. The nano-pyramidal geometry on GaN nanorod which the InGan/GaN QWs are placed promotes significantly greater strain relaxation in the InGaN layer and combines the epitaxial lateral overgrowth (ELO) technique to eliminate the TDs. And the polarization-induced internal electric field (IEF) of InGaN/GaN MQWs on semipolar $\{10\bar{1}1\}$ surface is much smaller than on c-plane with the same Indium composition in the QWs [7]. In this work, we report that high crystalline and efficiency green emission $\{10\bar{1}1\}$ InGaN/GaN MQWs had been grown on the $\{10\bar{1}1\}$ facets of GaN nanopyramid arrays by selective area epitaxy (SAE) technique.

Experiments

A 2 μm thick GaN film grown on sapphire was processed into GaN nanorods firstly by nano imprint lithography (NIL). As can be seen in the Fig. 1(a), well aligned cylindrical shape GaN nanorods have been fabricated after NIL dry etching process. The diameter and the periodicity of GaN nanorod are 350 nm and 750 nm, respectively. Then, the side wall of GaN nanorods were passivated with 30 nm thick dielectric material by spin on glass (SOG) technique and the dry etching process was applied to make sure the appearance of the top of GaN nanorods. The processes of SOG had not been optimized such as the concentration of sol-gel solution, curing temperature...etc, which result some irregular aspect and uncovered region of the passivated surface of GaN nanorods as shown in the Fig. 1(b). Finally, the GaN nano pyramids and 10 pairs $In_{0.3}Ga_{0.7}N$(3 nm)/GaN(8nm) MQWs were selectively grown on the top of the GaN nanorods in sequence by AXITRON 2000HT metalorganic chemical vapor deposition (MOCVD) reactor, as illustrated in the schematic structure in Fig. 1(c). To make uniformly pyramidal growth, the GaN nano pyramid was grown at high growth pressure of 600 mbar and low temperature of 800℃ with low growth rate of about 0.5μm/hr [8]. From the scanning electron microscopy (SEM) images in Fig. 1(d), one can clearly see that the nano pyramid structures have been selectively grown on the top of GaN nanorods.

Results and discussion

Figure 2 shows the cross section TEM images of nanopyramid heterostructures. It can be found that the InGaN/GaN MQWs were only grown on the inclined planes which having inclination of 62° with c-plane. This demonstrated that semipolar $\{10\bar{1}1\}$ InGaN/GaN MQWs can be successfully grown by the SAE technique. Moreover, the threading dislocation (TD) was found to be bended due to the appearance of the inclined planes which change the propagation direction of TD during the ELO process, as indicated by the red dash line frame in Fig. 2 (a). Therefore, low TD, clear and sharp interfaces of the semipolar $\{10\bar{1}1\}$ InGaN/GaN MQWs was achieved, as shown in the high-magnification TEM image in Fig. 2(b).

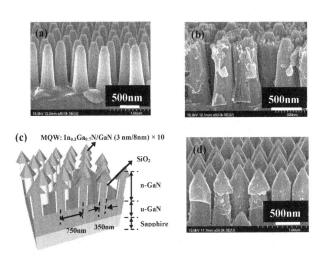

Figure 1. SEM images and the structure illustration figure. (a), GaN nanorod prepared by NIL. (b), side wall passivated by dielectric material. (c), Structure illustration. (d), after regrowing the GaN nanopyramid and InGaN/GaN MQWs.

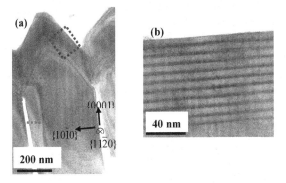

Figure 2. The cross section bright field TEM images of (a) the nanopyramids heterostructure, (b) the magnification image of $\{10\bar{1}1\}$ InGaN/GaN MQWs marked by blue dash line frame.

To study the effect of polarization field on the MQWs, the power photoluminescence measurements (PDPL) were performed by frequency doubled femto-second-pulse Ti:sapphire laser at a wavelength of 400 nm [9]. The excitation power was changed from 5×10^{-3} mW to 80 mW, which corresponds to the injection carrier density from 5×10^{16} cm^{-3} to 2×10^{19} cm^{-3}. Figure 3 summarizes the emission energies for $\{10\bar{1}1\}$ and (0001) InGaN/GaN MQWs under different excitation power. The emission energies at an excitation power of 5×10^{-3} mW are 2.413 eV (513.8 nm) and 2.351 eV (527.4 nm) for $\{10\bar{1}1\}$ and (0001) MQWs, respectively. When increasing the excitation power, the emission energy of both MQW samples increases, which results from the carrier screening effect against the IEF. But it is worth noting that the blueshift in $\{10\bar{1}1\}$ MQWs is much smaller than that in (0001) MQWs. These results indicate that the QCSE in $\{10\bar{1}1\}$ MQWs is suppressed. Furthermore, by the shifting of peak energy under various injection levels by theoretical model at RT [10], the IEF is found to be remarkably reduced from 1.7 MV/cm to 0.5 MV/cm as changing the growth plane from (0001) plane to $\{10\bar{1}1\}$ plane.

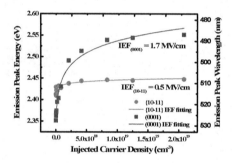

Figure 3. PL peak energy versus injected carrier density at room temperature by power dependent measurements of the $\{10\bar{1}1\}$ MQWs and (0001) MQWs.

Generally, reduction of the IEF facilitate a better overlap of electron and hole wave function that could result in the improvement of recombination rate of carriers. To verify the impact of the reduction of the IEF in green emission semipolar MQWs on the radiative recombination process, we further analyzed the carrier life time under injected carrier density 4×10^{18} cm^{-3} by time resolved PL (TRPL) at low temperature (LT) 15K and room temperature (RT) 300K. Figure 4 shows the TR spectra of the $\{10\bar{1}1\}$ and (0001) MQWs samples under injection carrier density 4×10^{18} cm^{-3}. As can be seen in Fig. 4, the fitted decay times for $\{10\bar{1}1\}$ MQWs at LT are approximately 0.22 and 0.09 ns, respectively, which both are nearly two order smaller than the values of 35.84 and 8.11 ns for (0001) MQWs. As has been reported that when the temperature is low enough to suppress the nonradiative, a purely radiative recombination at LT can be assumed [11]. Therefore, we can believe that the radiative lifetime in $\{10\bar{1}1\}$ MQWs has been significantly reduced (by nearly two orders magnitude) as a result of reduction of IEF, which leading to the increase of the emission efficiency. Furthermore, by using the equation $1/\tau_{PL}=1/\tau_r+1/\tau_{nr}$, the nonradiative lifetime at RT are estimated to be 0.15 ns and 10.48 ns for $\{10\bar{1}1\}$ and (0001) MQWs, respectively. The internal quantum efficiency (IQE) for $\{10\bar{1}1\}$ MQWs, calculated by the equation $\tau_{nr}(T)/[\tau_{nr}(T)+\tau_r(T)]$, is as high as 40.6%, which is much higher than that of 22.6% for (0001) MQWs. In addition, by carrying out the temperature dependent PL measurements, the IQE of $\{10\bar{1}1\}$ and (0001) MQWs were estimated to be about 38.1 % and 20.5 %, respectively, as shown in Table 1, which are consisted with the values obtained from TRPL. These results are encouraging for developing high In composition green emitters by using $\{10\bar{1}1\}$ semipolar surface as a growth plane to solve the efficiency green gap of III-nitride light emitters.

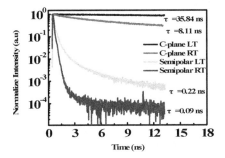

Figure 4. TRPL spectra of the $\{10\bar{1}1\}$MQWs and (0001) MQWs at 15 K and 300 K under injection carrier density 4×10^{18} cm^{-3}.

Table 1. The carrier lifetime and the fitting results of the TRPL measurement, and the IQE estimation by carrier lifetime.

	Plane	(0001)	$\{10\bar{1}1\}$
LT	τ_r(ns)	35.84	0.22
RT	τ_{PL}(ns)	8.11	0.09
	τ_{nr}(ns)	10.48	0.15
IQE@ 4×10^{18} cm^{-3}	By τ	22.6%	40.6%
	By PDPL	20.5%	38.1%

Conclusions

In conclusion, we successfully developed the high crystalline and high efficiency green emission semipolar $\{10\bar{1}1\}$ InGaN/GaN MQWs on the facets of GaN nano-pyramid arrays by SAE technique. The cross section TEM images demonstrated that low TD, clear and sharp interfaces of the semipolar $\{10\bar{1}1\}$ InGaN/GaN MQWs was achieved. The PDPL measurements indicated that the IEF of the $\{10\bar{1}1\}$ MQWs was remarkably reduced from 1.7 MV/cm to 0.5 MV/cm, as comparing with (0001) MQWs. As a result, the radiative lifetime in $\{10\bar{1}1\}$ MQWs has been significantly reduced (by nearly two orders magnitude), which increases the IQE by about 80%. These results evince that semipolar $\{10\bar{1}1\}$ plane is a great potential growth surface to develop high efficiency green light emitters and semiconductor laser diode for III-nitride light emitters.

Acknowledgments

The authors would like to thank Dr. T. C. Hsu and M. H. Shieh of Epistar Corporation for their technical support, and Professor Jihperng Leu of National Chiao Tung University for their assistance in dielectric material coating. This work was founded by the National Science Council in Taiwan under grant number, NSC NSC98-3114-M-009-001 and NSC99-2221-E-009-001.

Reference

[1] J. W, W. Walukiewicz, K. Yu, J. Ager, E. Haller, H. Lu, and H. Schaff, Appl. Phys. Lett. 80, 4741 (2002).

[2] P. T. Barletta, E. A. Berkman, B. F. Moody, N. A. El-Masry, A. M. Emara, M. J. Reed, and S. M. Bedair, Appl. Phys. Lett. 90, 151109 (2007).

[3] H. Masui, S. Nakamura, S. P. DenBaars, and U. K. Mishra, IEEE Trans. Electron Devices vol. 57, pp. 88, 2010.

[4] Y. J. Lee, S. Y. Lin, C. H. Chiu, T. C Lu, H. C. Kuo, S. C. Wang, S. Chhajed, J. K. Kim, and E. F. Schurbert, Appl. Phys. Lett. 94, 141111 (2009).

[5] H.W. Lin, Y.J. Lu, H.Y. Chen, H.M. Lee, and S.J. Gwo, Appl. Phys. Lett. 97, 073101 (2010).

[6] H. Sekiguchi, K. Kishino, and A. Kikuchi, Appl. Phys. Lett. 96, 231104 (2010).

[7] G. T. Chen, S. P. Chang, J. I. Chyi and M. N. CHang, Appl. Phys. Lett. 92, 241904 (2008).

[8] K. Hiramatsu, K. Nishiyama, M. Onishi, H. Mizutani, A. Motogaito, H. Miyake, Y. Iyechika, T. Maeda, J. Crystal Growth 221, 316 (2000).

[9] Y. J. Lee, C. H. Chiu, C. C. Ke, P. C. Lin, T. C. Lu, H. C. Kuo, and S. C. Wang, IEEE J. Sel. Top. Quantum Electron vol. 15, No. 4, pp. 1137-1143 (2009).

[10] A. Chtanov, T. Baars, and M. Gal, Phys. Rev. B 53, 4704 (1996)

[11] A. Sasaki, S. I. Shibakawa, Y. Kawakami, K. Nishizuka, Y. Narukawa, and T, Mukai, Jpn. J. Appl. Phys. 45, 8719 (2006)..

CdTe/CdS

Mater. Res. Soc. Symp. Proc. Vol. 1324 © 2011 Materials Research Society
DOI: 10.1557/opl.2011.1148

Morphology control of copper indium disulfide nanocrystals

Marta Kruszynska, Holger Borchert, Jürgen Parisi and Joanna Kolny-Olesiak

University of Oldenburg, Department of Physics, Energy and Semiconductor Research
Laboratory, Carl-von-Ossietzky-Str. 9-11, 26129 Oldenburg, Germany

ABSTRACT

In this report, we present a hot-injection strategy for the synthesis of $CuInS_2$ (CIS) nanocrystals with hexagonal, pyramidal and nanorod shapes. For that purpose copper (I) and indium (III) acetates were dissolved in oleylamine as a high-boiling solvent. *Tert*-dodecanethiol (*t*-DDT) was used as a sulfur source. It was mixed with *1*-dodecanothiol (*1*-DDT) and injected at a high temperature. The presence of the second dodecanethiol was necessary to control the growth of the synthesized nanocrystals. We observed a strong influence of the *t*-DDT amount on the morphology of the CIS nanocrystals. By the variation of the injected solution uniform CIS nanorods with different aspect ratio and size were obtained.

INTRODUCTION

A number of studies have been already reported on the synthesis of monodispersed binary chalcogenide semiconductor nanocrystals, while much less attention is focused on the ternary I-III-VI$_2$ family, e.g. $CuInS_2$ (CIS) nanocrystals [1-4]. This material is of great interests for photovoltaic applications due to the high absorption coefficient ($\sim 10^5$ cm^{-1}) in the visible region and radiation stability, a band gap of ~ 1.5 eV, and a relatively low toxicity [5-8]. CIS nanocrystals can be a promising candidate for organic-inorganic hybrid solar cells, since recently a successful charge transfer was observed in CIS/polymer blends [9]. Apart from photovoltaic applications, CIS nanoparticles can, e.g., be used as a fluorescence marker for in vivo biological imaging [10]. Various routes for the synthesis of CIS nanoparticles were presented in the literature. CIS nanocrystals have been prepared by a solvothermal method as well as a single source decomposition and hot-injections technique [12-13]. Depending on the reaction conditions $CuInS_2$ nanoparticles can have different crystallographic structures: chalcopyrite, wurtzite or zincblende [14-17]. Although, CIS nanocrystals with different morphology, e.g. nanotubes, hexagonal platelets and quasi-spherical particles, could be synthesized, it is still a challenge to obtain nanocrystals with a narrow size and shape distribution. There are only few reports about a successful synthesis of monodisperse CIS nanocrystals [6, 18]. In some cases, the synthesis leads to the formation of Cu_2S-In_2S_3 heterostructured nanocrystals [19].

In this report, we present a synthetic strategy to obtain $CuInS_2$ nanocrystals with controllable shape and size. We use a method based on our previous work [18] and study the influence of the composition of the injection solution, containing a mixture of *1*-dodecanothiol and *tert*-dodecanothiol, on the shape of the resulting nanocrystals. The structure, morphology and composition of the final materials were characterized by powder x-ray diffraction (XRD), transmission and scanning electron microscopy (TEM and SEM) and energy dispersive x-ray (EDX) analysis. The optical properties of $CuInS_2$ nanocrystals were investigated with UV-vis absorption spectroscopy.

EXPERIMENT

Materials. Copper (I) acetate (CuAc, 97 %), trioctylphosphine oxide (TOPO, 90 %), *l*-dodecanethiol (*l*-DDT, 98+%), and *tert*-dodecanethiol (*t*-DDT, 98.5 %, mixture of isomers) were obtained from Aldrich. Indium (III) acetate (InAc3, 99.99 %, metal basis) was purchased from Alfa Aesar and oleylamine (OLAM, approx. C18-content 80-90 %) from Acros.
Synthesis of CuInS$_2$ Nanocrystals. In a typical reaction, CuAc (0.5 mmol), InAc$_3$ (0.5 mmol) and TOPO (1.5 mmol) were mixed with OLAM (10 ml) under vacuum at room temperature. After 30 min, the reaction system was turned to nitrogen atmosphere and heated to 240 °C. During heating, the solution changed the color from green through dark blue and turbid green to yellow/brown, and at that moment a mixture of *l*-dodecanethiol/*tert*-dodecanethiol was rapidly injected. The process of CIS nanocrystals formation was stopped after 1 h. In all experiments, the concentration of Cu and In was fixed and the ratio between *l*-DDT and *t*-DDT was varied as is presented in Tab.1. The final materials were washed several times with ethanol and redispersed in chlorobenzene for further characterization.

Table I. Variation of the content of the injected solution.

Volume ratio between *l*-DDT and *t*-DDT	*l*-DDT (ml)	*t*-DDT (ml)	*l*-DDT (%)	*t*-DDT (%)
50:1	12.5	0.25	98	2
10:1	2.5	0.25	91	9
5:1	1.25	0.25	83	17
1:1	0.25	0.25	50	50
1:5	0.25	1.25	17	83
1:7	0.25	1.75	12	88
1:10	0.25	2.5	9	91
1:50	0.25	12.5	2	98
1:0	0.25	0	100	0
0:1	0	0.25	0	100

Characterization methods. TEM images were taken with a Zeiss EM 902A microscope (80 kV acceleration voltage). UV-vis absorption spectra were measured on a Varian Cary 100 Scan spectrophotometer. Powder X-ray diffraction was measured with a PANalytical X'Pert PRO MPD diffractometer using Cu K$_\alpha$ radiation and Bragg-Brentano θ-2θ geometry. SEM images and stoichiometry of nanocrystals were measured with a FEI **Helios NanoLab 600i DualBeam microscope.**

DISCUSSION

CIS nanoparticles were obtained by a reaction between copper (I) acetate with indium (III) acetate and *tert*-dodecanothiol in oleylamine as solvent. The sulfur precursor was injected at a high temperature, and in this way the nucleation of the nanoparticles was precisely controlled. We use a combination of two thiols: *tert*-dodecanethiol and *l*-dodecanethiol. *Tert*-dodecanethiol decomposes more easily than 1-dodecanthiol, because of its structure (more stable decomposition products), thus it serves as the main sulfur source. *1*-Dodecanethiol is more stable under the reaction conditions. That is why it mainly controls the reactivity of the copper monomers and stabilizes the nanocrystals. Additionally, nanocrystals were stabilized with trioctylphosphine oxide. During the reaction, the solution changed the color from clear yellow through orange to deep brown, and this behavior can indicate on the successful growth of $CuInS_2$ nanocrystals. We investigated the influence of the amount of *1*-DDT and *t*-DDT on the $CuInS_2$ nanocrystals formation.

We have considered a large number of samples containing various amounts of dodecanethiols, but only selected samples are presented here in detail. The morphology of synthesized nanocrystals was investigated with TEM and SEM. Fig. 1a-b and Fig. 2a present CIS nanocrystals synthesized with a small amount of *t*-DDT (the injection solution contains 9, 17 and 2 % of *t*-DDT, respectively).

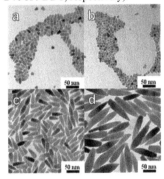

Figure 1. TEM images of $CuInS_2$ nanocrystals with (a) 10:1, (b) 5:1, (c) 1:5, and (d) 1:10 volume ratio between *1*-DDT:*t*-DDT.

The nanocrystals have a triangular shape and they are smaller than the nanoparticles synthesized with larger *t*-DDT amounts. The presence of a large amount of *1*-DDT lowers the reactivity of the monomers, which can delay the nucleation process. Not all the nanoparticles are formed at the same time and some dispersion in size is observed. When the amount of the sulfur source is increased to 83, 88 and 91 %, uniform nanorods are formed (Fig. 1c-d and Fig. 2b-c).

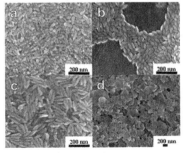

Figure 2. SEM images of CuInS$_2$ nanocrystals with (a) 50:1, (b) 1:7, (c) 1:10, and (d) 0:1 volume ratio between l-DDT:t-DDT.

This behavior can be explained by the increased chemical potential of the solution containing a large amount of t-DDT, which facilitates the formation of elongated shapes. The length and the width of the nanoparticles can be precisely control by changing the amount of t-DDT. At 83 % of t-DDT in the injected solution, the average length and width of the nanocrystals are 37 nm and 12.3 nm, respectively (aspect ratio of 3) (Fig. 1c). When the amount of t-DDT increases to 88 %, the aspect ratio decreases to 2.25. However, when l-DDT is completely excluded from the injected solution, the growth of hexagonal nanoplates with sizes between 203 and 564 nm is observed (Fig. 2d).

The crystallographic structure of CuInS$_2$ nanocrystals was studied by XRD. Fig. 3 shows the patterns of the samples described above. Independently on the t-DDT content, only wurtzite structure was observed. The variations in the relative intensities of the Bragg reflections between the patterns can be explained by different morphologies of the nanocrystals as well as by texture effects. As can be seen from the SEM and TEM images, the nanocrystals are not randomly oriented on the substrate.

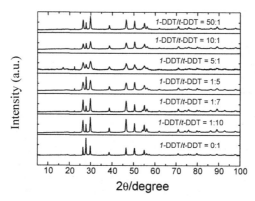

Figure 3. XRD patterns of CuInS$_2$ nanocrystals synthesized with various volume ratio between l-DDT:t-DDT. The patterns correspond to the wurtzite structure [18].

The chemical composition of nanocrystals was evaluated by EDX spectroscopy. The composition of the colloidally prepared nanocrystals does not change strongly, when the amount of t-DDT is varied. The ratio between Cu:In:S was found to be close to the 1:1:2 ratio in all cases.

The absorption spectra of CIS nanocrystals with various sizes and shapes are presented in Fig. 4. The nanocrystals absorb in the whole visible range. Due to the size quantization effect the absorption spectra of smaller particles are blue shifted compared with the band gap of the bulk material. A bandgap of 1.38 eV can be calculated from the absorption onset of the samples with the largest sizes. This value is smaller than the bandgap of bulk CIS with chalcopyrite structure (1.53 eV) [20]. This might be related to the crystal structure of the particles which is the wurtzite structure, i.e. a disordered polymorph of the chalcopyrite structure with the cation sublattice randomly occupied by Cu and In inons [20]. Theoretical calculations show that disordering of cations in the chalcopyrite structure induces a band gap decrease [21]. A bandgap of 1.1 eV has already been reported for nanocrystals with wurtzite structure [20].

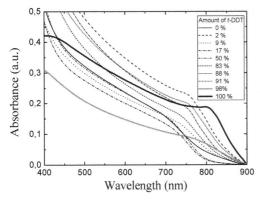

Figure 4. UV-vis absorption spectra of $CuInS_2$ nanocrystals with various amount of *t*-DDT.

CONCLUSIONS

We have investigated the possibility to control the shape of $CuInS_2$ nanocrystals by changing the composition of the injection solution, containing *tert*-dodecanethiol as sulfur source and *1*-dodecanethiol as stabilizer. Our results show, that the ratio between the two thiols has an influence on the final shape and size of $CuInS_2$ nanocrystals. The morphology of nanocrystals can be varied from hexagonal plates, through nanorods to pyramidal shapes. The nanocrystals have all a wurtzite crystal structure. We found out, that an excess of the *t*-DDT is necessary to obtain nanocrystals with a narrow size and shape distribution, however *1*-DDT cannot be completely excluded from the injected solution, because in its absence CIS nanocrystals with a broad size distribution are formed.

ACKNOWLEDGMENTS

This work was financially supported by the EWE Research Group "Thin Film Photovoltaics" as well as the Federal Ministry of Education and Research (BMMF, project number 03SF0338C).

REFERENCES

1. C. B. Murray, D. J. Norris and M. G. Bawendi, *J. Am. Chem. Soc.* **115**, 8706 (1993).
2. A.P. Alivisatos, *Science* **271**, 933 (1996).
3. Z.A. Peng and X. Peng, *J. Am. Chem. Soc.* **123**, 183 (2001).

4. H. Du, C. Chen, R. Krishnan, T.D. Krauss, J.M. Harbold, F.W. Wise, M.G. Thomas and J. Silcox, *Nano Lett.* **2**, 1321 (2002) .
5. F.M. Courtel, A. Hammami, R. Imbeault, G. Hersant, R.W. Paynter, B. Marsan and M. Morin, *Chem. Mater.* **22**, 3752 (2010).
6. D. Pan, L. An, Z. Sun, W. Hou, Y. Yang, Z. Yang and Y. Lu, *J. Am. Chem. Soc.* **130**, 5620 (2008).
7. R. Klenk, U. Blieske, V. Dieterle, K. Ellmer, S. Fiechter, I. Hengel, A. Jäger-Waldau, T. Kampschulte, C. Kaufmann, J. Klaer, M.C. Lux-Steiner D. Hariskos, M. Ruckh and H.W. Schock, *Sol. Energ. Mat. Sol. C.* **49**, 349 (1997).
8. B. Tell, J.L. Shay, H.M. Kasper, *Phys. Rev. B* **4**, 2463 (1971).
9. M. Kruszynska, M. Knipper, J. Kolny-Olesiak, H. Borchert and J. Parisi, *Thin Solid Films* (in press).
10. L. Li, T. J. Daou, I. Texier, T. T. Kim Chi, N.Q. Liem and P. Reiss, *Chem. Mater.* **21**, 2422 (2009).
11. R. Xie, M. Rutheford and X. Peng, *J. Am. Chem. Soc.* **131**, 5691 (2009).
12. J.P. Xiao, Y. Xie, R. Tong and Y.T. Qian, *J. Solid State Chem.* **161**, 179 (2001).
13. S.L. Castro, S.G. Bailey, R.P. Raffaelle, K.K. Banger and A.F. Hepp, *J. Phys. Chem. B.* **108**, 12429 (2004).
14. J.J.M. Binsma, L.J. Giling, and J. Bloem, *J. Cryst. Growth* **50**, 429 (1980).
15. K. Nose, Y. Soma, T. Omata and S. Otsuka-Yao-Matsuo, *Chem. Mater.* **13**, 2607 (2009).
16. S.K. Batabyal, L. Tian, N. Venkatram, W. Ji and J.J. Vittal, *J. Phys. Chem. C* **113**, 15037 (2009).
17. D. Pan, L. An, Z. Sun, W. Hou, Y. Yang, Z. Yang and Y. Lu, *J. Am. Chem. Soc.* **130**, 5620 (2008).
18. M. Kruszynska, H. Borchert, J. Parisi and J. Kolny-Olesiak, *J. Am. Chem. Soc.* **132**, 15976 (2010).
19. S.H. Choi, E.G. Kim and T. Hyeon, *J. Am. Chem. Soc.* **128**, 2520 (2006).
20. Y. Qi, Q. Liu, K. Tang, Z. Liang, Z. Ren and X. Liu, *J. Phys. Chem. C* **113**, 3939 (2009).
21. C. Rincon, *Phys. Rev. B* **45**, 12716 (1992).

Mater. Res. Soc. Symp. Proc. Vol. 1324 © 2011 Materials Research Society
DOI: 10.1557/opl.2011.1345

A Comparative Study of the Thin-Film CdTe Solar Cells with ZnSe/TCO and the Cds/TCO Buffer Layers

Tamara Potlog[1], Nicolae Spalatu[1]
Arvo Mere[2], Jaan Hiie[2], Valdek Mikli[2]
[1]Physics Department, Moldova State University, A. Mateevici str. 60, Chisinau, MD 2009 Republic of Moldova
[2]Department of Materials Science, Tallinn University of Technology, Ehitajate tee 5, Tallinn, 19086 Estonia

ABSTRACT

The growth of ZnSe and CdTe thin films by close spaced sublimation is examined. The investigations show that ZnSe films deposited on glass substrates are polycrystalline and exhibit wurtzite-zinc-blende polytypism. The CdTe films grown on glass/SnO_2/ZnSe are polycrystalline and have an f.c.c. zinc-blende structure as in the case of a glass/SnO_2/CdS buffer layer. The electric and photovoltaic parameters of ZnSe/CdTe solar cells depend on the ZnSe film thickness. Furthermore, it is shown for the first time that the best photovoltaic parameters are achieved using a Zn buffer layer at the interface between ZnSe and CdTe.

INTRODUCTION

The investigation of different types of semiconductor heterojunctions for easy photovoltaic solar energy conversion with simple fabrication and low cost has assumed special significance in recent years. As a result, many heterojunction solar cells are currently being explored. The solar cells based on CdTe material received wide attention. CdTe has a large optical absorption coefficient ($>10^4$ cm^{-1}). A small amount of CdTe (2-8 μm thick) is needed for the absorber layer (100 times thinner than typical crystalline-Si solar cells). For example, until now, the high efficiency (16.5%) of CdTe based solar cells has been achieved using a CdS/CdTe heterojunction [1]. However, the band gap energy of CdS (2.4 eV) is relatively small for the CdS/CdTe heterojunction solar cells, since 0.1 μm of a CdS film absorbs 36% of the incident radiation with an energy higher than 2.42 eV. Therefore, it is widely thought that it is possible to improve the efficiency of solar cells based on CdTe using alternative window materials. One of them is ZnSe. It has a larger bandgap (E_g =2.7 eV) than CdS and should allow more light to pass without being absorbed. Researchers [2] calculated the lattice mismatch using the electron affinity of the semiconductors and found CdTe/ZnSe to be 12.5%. The incorporation of S into ZnSe will form $ZnSe_xSe_{1-x}$ with a wider energy band gap. The previous paper [3] indicates that the contact potential of the formed junction between $ZnSe_xSe_{1-x}$ and CdTe is almost the highest among the II-VI alloys/compounds available to make window layers; hence, this combination has an excellent potential. The aim of this work is to grow a ZnSe thin film layer to be used with CdTe absorber layers. This layer will be deposited utilizing the same close sublimation (CSS) method as in the case of CdS/CdTe thin film cells [4]. The optimal technological conditions will be determined and correlated to the formation of ZnSe/CdTe devices with good photovoltaic parameters. The optimized high-efficiency ZnSe/CdTe photovoltaic devices will be compared with the best CdS/CdTe thin film cells prepared at Moldova State University.

EXPERIMENTAL DETAILS

Thin films of ZnSe, CdS and CdTe for heterojunctions were obtained by the CSS technique developed at the MSU [5]. We prepared the two sets of untreated and treated glass/SnO$_2$/ZnSe/CdTe and glass/SnO$_2$/ZnSe/Zn/CdTe samples in a saturated solution of CdCl$_2$. It is well known that the CdCl$_2$ treatment is an important step in the formation of a solar cell containing CdTe. The obtained structures were dipped into a CdCl$_2$ saturated solution for 30-35 min and then annealed in the atmosphere at 410°C. In our previous works [4-6] the optimal growth conditions for CdTe for high-performance CdS/CdTe solar cells were T$_s$=340°C substrate temperature and T$_{ev}$=620°C evaporator temperature. This technological regime was also used for the preparation of ZnSe/CdTe and ZnSe/Zn/CdTe heterosystems.The upper CdTe thin films were deposited directly on polycrystalline sublayers of CdS and ZnSe or ZnSe/Zn deposited on glass/SnO$_2$ substrates. The ZnSe layers were deposited at T$_s$ = 450°C and T$_{ev}$ = 900°C. Also, the cleaned glass slides were used as substrates for XRD study of ZnSe thin films. This attempt is made due to a lack of knowledge and understanding of the fundamental structural properties of ZnSe application in photovoltaic devices. Structural investigations of the films were performed using a Rigaku X-ray diffractometer with Cu/40 kV/40 mA radiation (λ = 1.54056 Å), a scan speed/ duration time -5.0000 deg/min, a K-beta filter, diffraction angles in the range of 20°≤2θ≤ 90° (where 2θ is the Bragg angle). The XRD analysis was performed using Rigaku software PDXL. The morphology of the surface and cross section were studied using a scanning electron microscope. The study of the current-voltage characteristics of solar cells in the dark and with 100 mW/cm^2 illumination was carried out according to the standard methods.

DISCUSSION
Structure and substructure analysis

The SEM images show that the ZnSe thin film deposited for 4 min has a granular structure and with an increase in the time of deposition (11 min) the grain sizes of crystallites increase and the cross-sectional images indicate that the ZnSe layer grows in a columnar morphology. The ZnSe thin films deposited on glass substrates were marked as 1-3. The XRD spectra of the ZnSe thin films for three different timeframes of deposition are illustrated in Fig 1. The structure of (1) and (2) films are composed of a mixture of cubic (zincblende) and hexagonal (wurtzite) ZnSe crystalline phases. However, the diffracted intensity along the (111) cubic peak in the (2) ZnSe sample is 30 times higher than the diffracted intensity along the same cubic peak distinguished in the (1) ZnSe sample. The increase in film texture with increasing time of deposition is probably due to the increase of the diffusion velocity of atoms on the substrate, which allows deposition to occur easily on the stable (111) plane. From the point of view of the growth mechanism along the [111] cubic preferential direction the hexagonal phase can be considered as a stacking fault of the cubic phase and vice versa. Such metastable hexagonal phase can be promoted to the presumably more stable cubic phase by the process of annealing [7]. The films prepared for a longer time of deposition have only a cubic structure. The lattice parameter of the cubic structure estimated from the position of the peaks of all ZnSe films corresponds to the value of $a = b=c = 5.676$ Å. The cubic phase in all samples corresponds to the

216: F-43m space group. The wurtzite phase corresponds to the 186: 63mc space group. For the cubic phase, using the Williamson-Hall method, it was possible to estimate the average grain sizes for (1), (2) and (3) ZnSe samples to be 593 Å, 687 Å and 712 Å, respectively.

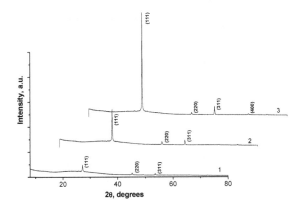

Figure 1. XRD pattern of the ZnSe films with the time of deposition, min:(1)-4; (2)-8; (3)-11.

Figure 2 shows the diffraction pattern for untreated and treated CdTe thin films deposited on glass/SnO$_2$/ZnSe and glass/SnO$_2$/ZnSe/Zn substrates. CdTe thin films deposited on glass/SnO$_2$/ZnSe/Zn were marked as Z1Z and Z2Z; those deposited on glass/SnO$_2$/ZnSe were marked as Z3 and Z4. In all the cases the ZnSe thin film was used as a window layer with the deposition time of 11 min. As one can see from Fig.2 all samples produced diffraction data from the CdTe layer that consisted of a high intensive peak at 2θ = 23.7° which corresponds to the (111) crystallographic plane indicating a highly oriented film. Also, in all diffraction data the peaks for cubic CdTe and SnO$_2$ (not marked in the diffraction spectra from Fig.2)) were easily identified. This shows that the CdTe films were grown with a cubic structure and the crystals grow preferentially along the [111] direction. The qualitative analysis of the XRD pattern of the glass/SnO$_2$/ZnSe/Zn/CdTe treated structure indicates that in this case at 2θ =23.7° both phases of the cadmoselite (CdSe) and of the CdTe appear. The calculated value of the lattice parameter a = 6.488 Å for the untreated and treated CdTe deposited on glass/SnO$_2$/ZnSe and the much higher values a = 6.492 Å for the untreated and treated CdTe films deposited on glass/SnO$_2$/ZnSe/Zn substrates than the one for a powder sample (6.481 Å) is an indication that the stress in these films is of a compressive nature.

Dark current-voltage study

The investigation of the dark current-voltage characteristics of several ZnSe/CdTe solar cells with different thicknesses of ZnSe indicate that with the increasing thickness of the ZnSe film the direct curves of the I-U characteristic shift in the direction of the current abscissa, which corresponds to an increase in the build-in voltage from 0.46 to 0.52 V. A further increase in the thickness of the ZnSe layer shows that build-in voltage remains constant. Thus, this thickness of ZnSe was taken as a value of the penetration depth of the contact field in the p/CdTe region of the heterostructure. The rectification coefficient of the heterostructure is about 2 for the ZnSe thin film deposited for 4 min and increases to 60 for the ZnSe thin film deposited for 11 min. A further increase in the thickness of the ZnSe layer results in an increase in the series resistance of the heterojunction, which limits the direct current that leads to a decrease in the rectification coefficient.

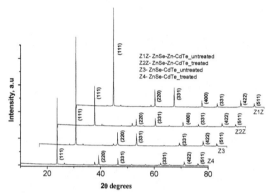

Figure 2. XRD pattern of the CdTe films deposited on glass/SnO$_2$/ZnSe substrates treated and untreated in CdCl$_2$, with and without a Zn buffer layer.

The effect of a window buffer layer on the illuminated current-voltage characteristics

The structure and interface play an important role in the performance of the CdTe heterojunction solar cell. The photovoltaic characteristics were investigated under the illumination of the heterostructures through wide band gap semiconductors, i.e. CdS and ZnSe. It is evident from Fig. 3 (a-f, top view) that the morphology of CdTe depends on the structure, morphology of the sublayer and chlorine treatment. For comparison, Fig. 3 (a, b) shows the morphology of CdTe with a thickness of about 12 μm deposited on 0.4-μm CdS films before and after CdCl$_2$ treatment. It is interesting to note that the grain sizes in the thin CdS film are small (~0.1 μm). The morphology of CdTe deposited on CdS at the above substrate temperature is characterized by a compact layer with large crystallites especially after CdCl$_2$ treatment (the average grain size is 5.9 μm). The impact of CdCl$_2$ treatment on cell operation is increased photocurrent and open-circuit voltage, and reduced shunting.

Table 1 Photovoltaic parameters of the solar cells based on CdTe thin films

Samples	J_{sc}, mA/cm^2	U_{oc}, V	FF, %	η, %	R_s, Ohm·cm^2	R_{sh}, Ohm·cm^2
ZnSe/Zn/CdTe	20.8	0.63	42	5.5	13.7	256
ZnSe/CdTe	10.7	0.686	38	2.8	38.6	193.4
CdS/CdTe	23,9	0,79	51	9,6	8.4	3769

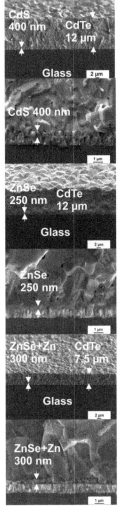

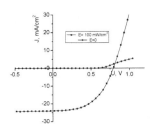

Figure 4. Current-voltage characteristics of the CdS/CdTe solar cells; 100 mW/cm^2, T = 300 K.

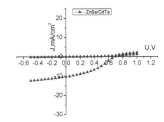

Figure 5. Current-voltage characteristics of the ZnSe/CdTe solar cells: 100 mW/cm^2, T = 300 K.

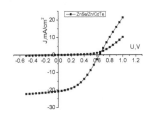

Figure 3. SEM images of treated (b, d, f) and untreated (a, c, e) CdTe films deposited on different sublayers. Left: top view; right-side view.

Figure 6. Current-voltage characteristics of the ZnSe/Zn/CdTe solar cells: 100 mW/cm^2, T = 300 K.

The photovoltaic parameters of one of the best CdS/CdTe solar cells calculated at 100 mW/cm^2 current-voltage and T = 300 K (Fig. 4) are presented in Table 1. The value of the grain sizes of CdTe grown on ZnSe decreases from 4.7 μm to 4.0 μm before treatment and from 5.9

µm to 4.7 µm after $CdCl_2$ treatment, although the thickness of CdTe is the same as in the case of CdS/CdTe solar cells (Figs. 3 (a-d)). The current-voltage characteristics of the ZnSe/CdTe solar cells with ~ 4 µm-thicks ZnSe are shown in Fig. 5. The efficiency reaches about 2.8%. The heterojunction solar cells with thinner ZnSe films reduce all photovoltaic parameters. The relatively low open circuit voltage may be due to pinholes which allow shorting the paths formed during deposition. The series resistances are higher in all the prepared heterojunction solar cells and much higher in ZnSe/CdTe. ZnSe thin films exhibit a higher resistance than CdS. To reduce the resistance of ZnSe the buffer layer of Zn was evaporated at the interface between ZnSe and CdTe. The value of grain sizes decreases as evident from Fig. 3 (e, f) in comparison with Fig 3 (c, d). We assume that in this case the buffer layer of Zn decreases the porosity of the ZnSe film and reduces the effect of lattice mismatch between ZnSe and CdTe layers and, hence, improves the photovoltaic parameters of ZnSe/Zn/CdTe solar cells (Fig. 6.).

CONCLUSIONS

Firstly, the use of the CSS method for the growth of ZnSe layers allows obtaining ZnSe/CdTe thin-film solar cells with a fairly high current density. It should be noted that the photovoltaic parameters of ZnSe/CdTe solar cells depend on the thickness of the ZnSe film. An optimum thickness of the ZnSe film to obtain good rectification and high current density was found. Secondly, in this work we compared the photovoltaic parameters of the solar cells based on CdTe with $ZnSe/SnO_2$ and CdS/SnO_2 buffer layers prepared under the same conditions and we found that the solar cells with a CdS/SnO_2 buffer layer are the best.
Finally, the introduction of a Zn buffer layer between ZnSe and CdTe decreased the effect of lattice mismatch in the ZnSe/CdTe heterosystem and improved the efficiency.

ACKNOWLEDGMENTS

This research was supported by the EU 7th Framework Program PEOPLE International Research Staff Exchange Scheme project "Development of Flexible Single and Tandem II-VI Based High Efficiency Thin Film Solar Cells" GA-2008, No. 230861 and ESF G7608 grant.

REFERENCES

1. X. Wu, Solar Energy. 7, 803-814 (2004)
2. R.H. Bube, F.Buch, A.I. Fahrenbruch, Y.Y. Ma, and K. Mitchell, IEEE Transaction on Electron Devices, Vol. 24, 487-492 (1977).
3. A.L. Fahrenbruch and R.H.Bube. Fundamentals of Solar Cells. Publ. Academic Press (1983), pp.235-278.
4. L.Ghimpu, V.Ursaki, T.Potlog, and I. Tighineanu, Semiconductor Science and Technology, 20, 1127-1131 (2005).
5. T. Potlog G. Khrypunov. M. Kaelin, H. Zogg, and A. N. Tiwari. Thin-Film Compound Semiconductor Book, Volume 1012, 181-188, Photovoltaics-2007, edited by T. Gessert, K. Durose, C. Heske, S. Marsillac, and . Wada. www.mrs.org/meetings/.
6. T. Potlog, L Ghimpu, and C. Antoniuc. Thin Solid Film, 515, 5824-5827 (2007).
7. S. Jimenez-Sandoval, M. Melendez-Lira, I. Hernandez-Calderon, J. Appl. Phys. 72, pp. 4197-4202 (1992).

Mater. Res. Soc. Symp. Proc. Vol. 1324 © 2011 Materials Research Society
DOI: 10.1557/opl.2011.1057

Influence of Surface Preparation on Scanning Kelvin Probe Microscopy and Electron Backscatter Diffraction Analysis of Cross Sections of CdTe/CdS Solar Cells

H.R. Moutinho, R.G. Dhere, C.-S. Jiang, and M.M. Al-Jassim
National Renewable Energy Laboratory, Golden, CO 80401, USA

ABSTRACT

Electron backscatter diffraction (EBSD) provides information on the crystallographic structure of a sample, while scanning Kelvin probe microscopy (SKPM) provides information on its electrical properties. The advantage of these techniques is their high spatial resolution, which cannot be attained with any other techniques. However, because these techniques analyze the top layers of the sample, surface or cross section features directly influence the results of the measurements, and sample preparation is a main step in the analysis.

In this work we investigated different methods to prepare cross sections of CdTe/CdS solar cells for EBSD and SKPM analyses. We observed that procedures used to prepare surfaces for EBSD are not suitable to prepare cross sections, and we were able to develop a process using polishing and ion-beam milling. This process resulted in very good results and allowed us to reveal important aspects of the cross section of the CdTe films. For SKPM, polishing and a light ion-beam milling resulted in cross sections that provided good data. We were able to observe the depletion region on the CdTe film and the p-n junction as well as the interdiffusion layer between CdTe and CdS. However, preparing good-quality cross sections for SKPM is not a reproducible process, and artifacts are often observed.

INTRODUCTION

EBSD [1] is performed inside a scanning electron microscope (SEM), where the electrons from the beam are diffracted by the top layers of the material and collected by a detector positioned close to the sample surface. To increase the yield of diffracted electrons, the sample is tilted by 70°, which requires a flat surface to avoid surface features from preventing the electrons from reaching the EBSD detector (the shading effect). Furthermore, because the diffracted electrons come from the region close to the surface (about 20 nm deep), the quality of the surface is a key parameter for obtaining good EBSD data. In addition to conventional information, such as pole figures and inverse pole figures, EBSD provides unique information on orientation maps and boundaries' misorientation profiles, and it is the most reliable technique to provide surface grain size information.

SKPM [2] provides measurements of the electrical potential and electric field distribution on the sample surface. The technique provides maps of the surface potential simultaneously with atomic force microscopy (AFM) topographic images, which allows for the correlation between topography and electrical properties. As with EBSD, the advantage of SKPM over other techniques is the high spatial resolution. When applied to cross sections of biased CdTe/CdS solar cells, it reveals the location of the p-n junction and the distribution of the depletion region on the CdTe film and also allows for studying the interdiffusion layer between CdTe and CdS.

Because these techniques analyze the surface of the sample, sample preparation is a key step toward obtaining meaningful data. For instance, close-spaced sublimation (CSS) CdTe films are too rough to provide good EBSD data because of shading effects. Polishing the films produces a flat surface, but with poor quality, resulting in no Kikuchi patterns on the EBSD detector and, consequently, no EBSD data. In previous work [3], we found that good samples are produced by ion-beam milling, a combination of polishing and ion-beam milling, or polishing and etching in bromine methanol solution.

Although there are no shading effects in AFM, the maximum vertical range of the tip is about 7 μm, requiring that sample features be no taller than about 4 μm. However, because of convolution between topography and surface potential data, steps should be as flat as possible (no more than a few dozens of nanometers). Because of this, only samples that cleave provide good cross sections with minimal preparation. CdTe/CdS films, which are deposited on glass, require a polishing stage before any meaningful SKPM data can be obtained. After this, a light ion-beam milling stage can also be applied.

In this work we investigate procedures to prepare CdTe/CdS solar cell cross sections for EBSD and SKPM analysis, and report the information that can be obtained from the samples.

EXPERIMENTAL PROCEDURE

The samples used in this work had the following structure: Ag paste/Cu-doped graphite paste/CdTe/CdS/i-SnO_2/SnO_2/glass. The samples received a standard vapor $CdCl_2$ treatment at 400°C for 5 min.

The samples were polished in a MultiPrep system from Allied High Tech Products, Inc., using diamond lapping films with 30, 9, 3, 1, 0.5, 0.25 and 0.1-μm grits. Different polishing solutions and procedures were tried, but the results were similar. A last step using 0.05-μm alumina suspension was sometimes used, but again, the results of the analyses were similar. After polishing, some samples were etched in bromine/methanol solution for 2 s, while other samples were ion-beam milled on a Fischione system, model 1010 LAIMP. To avoid rounding the film during the polishing process, the samples were sandwiched using epoxy and glass slide. For the EBSD analysis, we used a conductive epoxy to diminish charging effects caused by the electron beam. For the SKPM analysis, because we wanted to analyze the sample under different bias conditions, we attached a wire to the back contact and used non-conductive epoxy to avoid short-circuiting the solar cell. It is important to mention that the procedures reported in this work may need to be modified for other materials and CdTe films deposited by other methods.

The EBSD analysis was performed in a SEM FEI Nova 630 NanoSEM using a EDAX Pegasus/Hikari A40 system. The SKPM measurements were performed in a ThermoMicroscope AutoProbe CP Research scanning probe microscope using Pt-Ir-coated Si tips.

RESULTS AND DISCUSSION

EBSD
We initially attempted to cleave (break) the samples, but no useful results were obtained due to shading between the several layers that were formed. Next, we tried to polish the sample and then etch using a bromine/methanol solution. This was the natural choice, because this process had been used successfully before to prepare the surface of CdTe films for EBSD [3], and a system for ion-milling cross sections was not available. The results are shown in Fig. 1. On

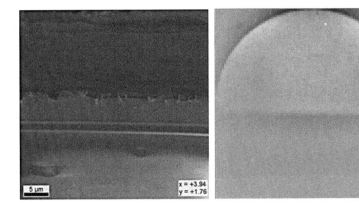

Fig. 1. Left: SEM image of the cross section of a CdTe/CdS cell after polishing and etching with bromine methanol solution. From the top: graphite paste/CdTe/CdS-SnO$_2$/glass. Right: Image on the EBSD detector at the location marked by a green x on the SEM image.

the SEM image we notice that the films are not as flat as we would expect. Also, there is a step between the graphite paste and the CdTe, and there is some residue on the surface of the CdTe film. The image on the detector shows a strong shading effect (the bottom half of the detector is dark, showing no detection of electrons) due to the step observed on the SEM image. On the top part of the detector, although there is no shading, there are no Kikuchi lines, indicating that the CdTe surface does not have good crystallinity. This is probably caused by the features on the CdTe film observed on the SEM image. We tried to solve this problem by changing the polishing process, but we were not successful. For every sample treated with bromine/methanol, there was some deposit on the CdTe film and a step between the back contact and the CdTe observed by SEM, and shading effects and no Kikuchi patterns observed on the EBSD detector. These experiments allowed us to conclude that bromine/methanol attacks the back contact and the epoxy, creating the step and leaving a residue layer on the CdTe, preventing EBSD data from being acquired.

To solve this problem, we were able to adapt an ion-beam milling used to prepare samples for TEM to mill our cross sections. The new cross sections were flat, without steps between the back contact and the CdTe, and excellent Kikuchi patterns were observed on the detector. However, we observed that, for strong ion milling, there was a small step between the CdTe and CdS-SnO$_2$ films, which would cause some shading close to this interface. To avoid this problem, we optimized the ion-beam milling process. Fig. 2 shows two inverse pole figure orientation maps of CdTe cross sections. For light ion milling (left) it is difficult to clearly see the film crystalline structure because only part of the damaged layer was removed. For intermediate milling conditions (not shown), although the crystalline structure could be easily observed, there was a lot of variation in the crystallographic orientation inside the grains. For the optimized milling conditions, the data was excellent (right). Using polishing followed by ion-beam milling, we are able to create good cross sections for EBSD on a routine basis.

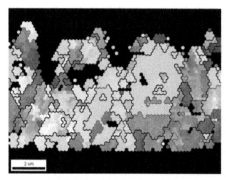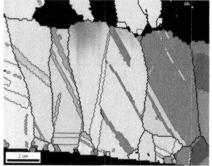

Fig. 2. EBSD of the cross section of a CdTe/CdS solar cell. Left: Light ion-beam milling (4 kV, 4 mA, 5°, 10 min). Right: Optimal ion-beam milling conditions (5 kV, 5 mA, 6°, 30 min). Dark lines denote grain boundaries, and red lines denote CSL Σ3 boundaries.

The analysis of the samples show that the CdTe film grows in a columnar way and has small grains that nucleate in the first stages of film growth at the interface with CdS. Analysis of pole figures shows that the film is randomly oriented. Almost every grain has few low-energy coincidence site lattice (CSL) Σ3 boundaries [4]. These boundaries are twins, generated by a rotation of 60° around a [111] crystallographic direction and have been observed before [5].

SKPM

For the SKPM measurements, to avoid the problems observed in the EBSD analysis, we did not use bromine/methanol etching. The best results were obtained by polishing or polishing followed by a light ion-beam milling, which provided very similar data. Some of the results are displayed in Fig. 3, which shows representative linescans of the potential (left) and electric field (right) on the cross section of CdTe/CdS cells. The figure on the left shows that there is a sharp decrease in the potential at the interface between CdTe and CdS. As the reverse bias is increased, there is an increase and spread in the potential inside the CdTe film, corresponding to the expansion of the depletion region with reverse bias. As expected, the potential drop inside the SnO_2 film is small. The figure on the right shows the electric field inside the device obtained by taking the first derivative of the potential linescans. There is a strong electric field at the junction, which is caused by the higher doping interdiffusion layer created during the formation of the device. The existence of this layer was confirmed by transmission electron microscopy (TEM) [6]. The position of the p-n junction coincides with the maximum of the electric field. The expansion of the depletion region on the CdTe with the increase in reverse bias is also observed in the figure. We applied Poisson's equation on the straight part of the electric field on the CdTe film on polished samples, and we were able to calculate the doping for several samples [6]. The calculated values varied from 1.1×10^{14} to 3.3×10^{14} cm^{-3}, which agrees well with values reported on the literature for CdTe.

Analyzing Fig. 3, we notice that the electric field at forward bias is negative, when it should be positive. This indicates the existence of an inversion layer on the CdTe/CdS surface,

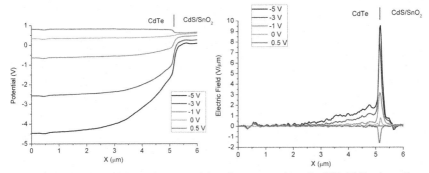

Fig. 3. Left: Linescans of the surface potential on the cross section of a CdTe/CdS solar cell after polishing and ligh ion-beam milling under different bias conditions (inset). Right: Corresponding linescans of the electric field.

which is caused by pinning of the Fermi level due to surface states. These surface states are affected by the sample preparation method, and only-polished samples present even stronger effects, such as a negative electric field at 0 bias. These results serve as a reminder that SKPM analyzes the surface of the sample, and care is needed when comparing the results with the bulk of the sample. Unfortunately, this is not the main problem. Fig. 4 shows linescans of the surface potential and electric field for another CdTe/CdS solar cell after polishing. The most striking feature on the surface potential linescans (left) is the crossing of the curves on the right (SnO_2 region). During the analysis, the SnO_2 film is grounded, and the reverse and forward biases are applied on the back contact. Because there is a drop in potential from where the SnO_2 is grounded to where the tip starts scanning, the linescans on the SnO_2 side are supposed to be slightly separated, as in Fig. 3. However, the theory cannot explain the behavior or the linescans on the SnO_2 side observed in Fig. 4. Furthermore, on the electric field linescans (right), we don't see a clear increase in the depletion region on the CdTe film as the reverse bias is increased, and we cannot explain the large electric field on the SnO_2 film, which is related to the strange shape of the potential signal on that region. We measured the efficiency parameters of the cell in Fig. 4, and V_{oc} before sample preparation and after analysis was around 800 mV, and efficiency was around 9.5%, which shows that the device was working during the measurements. We analyzed many samples and could not predict when a sample would behave like in Fig. 3 or Fig. 4. The conclusion is that we cannot reproduce the polishing/ion-beam milling process to the level required by SKPM. This problem makes it difficult to analyze unknown samples because it is difficult to separate results due to the sample properties from artifacts caused by the preparation process. Currently we are investigating new methods to prepare cross sections for SKPM.

CONCLUSIONS

Although etching with bromine/methanol solution produces good surfaces for EBSD analysis of CdTe, when applied to cross sections it attacks the film structure, resulting in large steps and contamination, and no useful EBSD data.

61

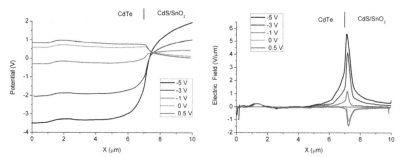

Fig. 4. Left: Linescans of the surface potential on the cross section of a CdTe/CdS solar cell after polishing under different bias conditions (inset). Right: Corresponding inescans of the electric field.

We developed a reliable procedure to prepare cross sections of CdTe/CdS solar cells for EBSD, which consists of polishing followed by ion-beam milling. An optimization of the milling parameters is important for the fabrication of good quality samples with just a small step between the SnO_2/CdS and CdTe films. Very good EBSD data can be routinely obtained, revealing the columnar character of the CdTe growth and details on the crystallographic structure of the film, such as grain size and boundary characteristics.

The best procedure to prepare cross sections of CdTe/CdS solar cells for SKPM was by polishing and polishing with a light ion-beam milling. Good SKPM data was obtained showing important aspects of the junction, such as the intermixed layer between CdTe and CdS, and the location of the p-n junction. However, the sample preparation process is not reliable, and the results are not always reproducible. Improvements in sample preparation will be needed before this technique can be used in a routine basis to study CdTe/CdS junctions.

ACKNOWLEDGEMENTS

This work was supported by the U.S. Department of Energy under Contract No. DE-AC36-08-GO28308 with the National Renewable Energy Laboratory.

REFERENCES
1. D. J. Dingley and D. P. Field, Mater. Sci. Technol. **13**, 69 (1997).
2. P. A. Rosenthal and E. T. Yu, J. Appl. Phys. **87**, 1937 (2000).
3. H. R. Moutinho, R. G. Dhere, M. J. Romero, C.-S. Jiang, B. To, and M. M. Al-Jassim, J. Vac. Sci. Technol. A **26**, 1068 (2008).
4. Y. Yan and M. M. Al-Jassim, J. Appl. Phys. **90**, 3952 (2001).
5. J. Quadros, A. L. Pinto, H. R. Moutinho, R. G. Dhere, and L. R. Cruz, J. Mater. Sci. **43**, 573 (2008).
6. H. R. Moutinho, R. G. Dhere, C.-S. Jiang, Y. Yan, D. S. Albin, and M. M. Al-Jassim, J. Appl. Phys. **108**, 074503 (2010).

Mater. Res. Soc. Symp. Proc. Vol. 1324 © 2011 Materials Research Society
DOI: 10.1557/opl.2011.1151

CdS_xTe_{1-x} Alloying in CdS/CdTe Solar Cells

Joel N. Duenow, Ramesh G. Dhere, Helio R. Moutinho, Bobby To, Joel W. Pankow, Darius Kuciauskas, and Timothy A. Gessert
National Renewable Energy Laboratory, 1617 Cole Blvd., Golden, CO 80401

ABSTRACT

A CdS_xTe_{1-x} layer forms by interdiffusion of CdS and CdTe during the fabrication of thin-film CdTe photovoltaic (PV) devices. The CdS_xTe_{1-x} layer is thought to be important because it relieves strain at the CdS/CdTe interface that would otherwise exist due to the 10% lattice mismatch between these two materials. Our previous work [1] has indicated that the electrical junction is located in this interdiffused CdS_xTe_{1-x} region. Further understanding, however, is essential to predict the role of this CdS_xTe_{1-x} layer in the operation of CdS/CdTe devices. In this study, CdS_xTe_{1-x} alloy films were deposited by radio-frequency magnetron sputtering and co-evaporation from CdTe and CdS sources. Both radio-frequency-magnetron-sputtered and co-evaporated CdS_xTe_{1-x} films of lower S content (x<0.3) have a cubic zincblende (ZB) structure akin to CdTe, whereas those of higher S content have a hexagonal wurtzite (WZ) structure like that of CdS. Films become less preferentially oriented as a result of a $CdCl_2$ heat treatment at ~400°C for 5 min. Films sputtered in a 1% O_2/Ar ambient are amorphous as deposited, but show CdTe ZB, CdS WZ, and CdTe oxide phases after a $CdCl_2$ heat treatment. Films sputtered in O_2 partial pressure have a much wider bandgap than expected. This may be explained by nanocrystalline size effects seen previously [2] for sputtered oxygenated CdS (CdS:O) films.

INTRODUCTION

A CdS_xTe_{1-x} layer forms by interdiffusion of CdS and CdTe during the fabrication of thin-film CdTe PV devices in the standard superstrate configuration. High-temperature processing steps such as the close-spaced sublimation of CdTe and the post-deposition $CdCl_2$ heat treatment (HT) contribute to formation of this alloy [1]. The CdS_xTe_{1-x} layer is thought to be important in fabricating high-performance CdTe devices because it relieves strain at the CdS/CdTe interface that would otherwise exist due to the large lattice mismatch (~10%) between these two materials. Our previous work indicated that the electrical junction is located in this interdiffused CdS_xTe_{1-x} region between a structurally compatible Te-rich n-type CdS_xTe_{1-x} alloy and p-type CdTe [1,3].

The CdS_xTe_{1-x} alloy has been found to follow Vegard's law [4], such that lattice parameter values obtained from X-ray diffraction measurements can be used to calculate the mole fraction (x) for different phases of CdS_xTe_{1-x}. The bandgap (BG) of CdS_xTe_{1-x} has been described as a quadratic function of x [5], with values decreasing below the CdTe BG value of 1.5 eV, to as low as 1.41 eV at x~0.3, before increasing at higher x values. The alloy system also exhibits a miscibility gap (two-phase region) in which both a CdTe-rich zincblende (ZB) phase and CdS-rich wurtzite (WZ) phase may be present simultaneously at equilibrium. The composition of each phase in the two-phase region is the same as in the corresponding single-phase regions at the edges of the miscibility gap, with the relative quantity of each phase present varying with x [6]. Single-phase films, however, have been grown within the miscibility gap, implying non-equilibrium growth methods were used. The phases generally separated after a $CdCl_2$ HT [6-8]. We found similar behavior in this study.

Further understanding of CdS_xTe_{1-x} alloys is essential to predict the role of the CdS_xTe_{1-x} layer in the operation of $CdS/CdTe$ devices. In this study, we investigate this alloy layer by depositing CdS_xTe_{1-x} films using two methods.

EXPERIMENTAL DETAILS

We deposited CdS_xTe_{1-x} films by radio-frequency (RF) magnetron sputtering using targets of three compositions: 10/90, 25/75, and 60/40 wt.% $CdS/CdTe$. Films were deposited at room temperature (RT; no intentional heating) and at 300°C. Because little difference was seen between films grown at RT and at 300°C, only the 300°C results are presented here. In addition, two deposition ambients were used—100% Ar and 1% O_2/Ar. The ratio was measured using an ion gauge. Films were also deposited by co-evaporation from CdTe and CdS sources using Radak II effusion cells. The geometry of the co-evaporation system enabled us to obtain a range of compositions during a single deposition. The films were deposited onto three different substrates—Corning 7059 glass, Corning 7059 glass/450-nm SnO_2:F/150-nm SnO_2, and Corning 7059 glass/SnO_2:F/SnO_2/125-nm sputtered CdS:O—to be amenable to the different types of characterization performed on the films. Reflectance and transmittance measurements were performed using a Cary 6000i UV-Vis-NIR spectrophotometer. Electron-probe microanalysis (EPMA) was performed using a beam energy of 5 kV to obtain film composition. X-ray diffraction (XRD; Rigaku Ultima IV) θ-2θ measurements were performed using Cu K_α radiation to examine the structure of the CdS_xTe_{1-x} films. Electron backscatter diffraction (EBSD; FEI FEG SEM Nova 630 NanoSEM with an EDAX Pegasus/Hikari A40 EDS/EBSD system) measurements were performed to examine grain orientation and size for selected films.

RESULTS AND DISCUSSION

Composition of the ~150-nm-thick sputtered CdS_xTe_{1-x} alloy films was measured using EPMA (Fig. 1, left panel). Films were grown in 100% Ar and 1% O_2/Ar both before and after a 5-min vapor $CdCl_2$ HT at 400°C. Films grown from the 10/90 wt.% $CdS/CdTe$ target show little compositional change between the different deposition ambients and after the $CdCl_2$ HT. Films

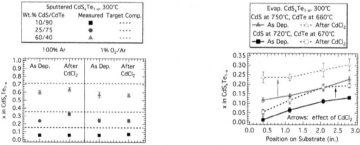

Fig. 1. Measured x values (obtained using EPMA) for sputtered (left) and evaporated (right) CdS_xTe_{1-x} films. Dashed lines in the left plot indicate the manufacturer's stated target composition. Evaporated films were measured at four positions because of the inherent composition gradient across the substrate due to the deposition system geometry.

grown from the 25/75 and 60/40 wt.% CdS/CdTe targets in the 100% Ar ambient display a slight enrichment in S (or, equivalently, a loss of Te) after the CdCl$_2$ HT. All films appear S-deficient, however, compared to the manufacturer's stated target composition. This may be due to differences in sticking coefficient on the substrate of the S and Te species. Two ~300-nm-thick evaporated CdS$_x$Te$_{1-x}$ films (Fig. 1, right panel) were deposited using different combinations of CdS and CdTe effusion cell temperatures. The as-deposited films (solid lines) show the expected gradient in x across the substrate. The evaporated films also show a significant relative enhancement in S (or loss of Te) after the CdCl$_2$ HT.

CdS$_x$Te$_{1-x}$ alloy film structure was investigated using XRD. Fig. 2 shows θ-2θ scans of films grown from the 10/90 wt.% CdS/CdTe target in the 100% Ar and 1% O$_2$/Ar ambients both before and after a CdCl$_2$ HT. Films examined using XRD were deposited on Corning 7059 glass/450-nm SnO$_2$:F/150-nm SnO$_2$/125-nm sputtered CdS:O substrates similar to those used for CdTe PV devices. Tetragonal SnO$_2$, cubic ZB CdTe, and hexagonal WZ CdS peaks are indicated. The as-deposited film grown in 100% Ar shows one prominent peak, CdTe ZB (111). Other peaks in this spectrum are due to the SnO$_2$ films on the substrate. After a CdCl$_2$ HT is performed on this film, the CdTe ZB (111) peak decreases in intensity, while small CdTe ZB (220) and (311) peaks, and CdS WZ (100), (002), and (101) peaks, appear. This behavior indicates a decrease in preferential orientation after the CdCl$_2$ HT. The film grown in 1% O$_2$/Ar is amorphous as deposited. After the CdCl$_2$ HT, however, many CdTe ZB and CdS WZ phases appear, in addition to prominent CdTe oxide phases (e.g., CdTeO$_3$, CdTe$_2$O$_5$). Further work is required to identify which oxide phases are dominant. A summary of XRD results for the sputtered films is shown in Table 1. Films grown in 100% Ar from the 60/40 wt.% CdS/CdTe target contain only CdS WZ phases, both before and after the CdCl$_2$ HT, whereas films of higher CdTe content in this ambient show both CdTe ZB and CdS WZ phases. All films grown in 1% O$_2$/Ar are amorphous as deposited. The CdTe oxide peaks observed after the CdCl$_2$ HT are much less intense for the 25/75 and 60/40 wt.% CdS/CdTe films than for the 10/90 wt.% film.

XRD measurements were also performed on evaporated CdS$_x$Te$_{1-x}$ alloy films both before and after a CdCl$_2$ HT at 390°C. Measurements were performed at several points on the film because of the composition gradient. The CdTe ZB (111) peak was observed to shift to an intermediate position between it and the adjacent CdS WZ (100) peak as the S content increased.

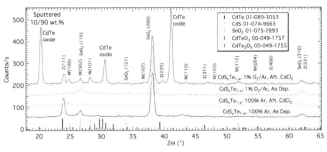

Fig. 2. XRD scans for sputtered CdS$_x$Te$_{1-x}$ films grown from the 10/90 wt.% CdS/CdTe target. Growth ambient and post-deposition treatment are shown on the plot. Peak positions from the powder diffraction files corresponding to these materials are shown at the bottom (with corresponding file numbers shown in the upper right).

Table 1. Phases present in sputtered CdS_xTe_{1-x} films grown in 100% Ar and 1% O_2/Ar before and after a $CdCl_2$ HT.

Wt.%	100% Ar		1% O_2/Ar	
CdS/CdTe	As Deposited	After CdCl₂	As Deposited	After CdCl₂
10/90	CdTe ZB	CdTe ZB CdS WZ	Amorphous	CdTe ZB CdS WZ Strong oxide phases
25/75	CdTe ZB Minor CdS WZ	CdTe ZB CdS WZ	Amorphous	CdTe ZB CdS WZ Oxide phases
60/40	CdS WZ	CdS WZ	Amorphous	CdS WZ Oxide phases

After the $CdCl_2$ HT, the CdTe ZB (111) peak shifts toward its expected position, while other CdTe ZB and CdS WZ phases appear, indicating a decrease in film preferential orientation. These data are summarized in Table 2. The separation of the CdTe ZB (111) and CdS WZ (100) peaks suggests that phase separation occurs as a result of the $CdCl_2$ HT.

Analysis was performed to extract the lattice constants and mole fraction of the CdTe ZB phase from the XRD data. The lattice constant a_{Cubic} was determined using

$$a_{Cubic} = d\sqrt{h^2 + k^2 + l^2},$$ where $d = \lambda_a/(2\sin\theta)$ is calculated from the measured 2θ peak position, $\lambda_\alpha = 1.540562$ Å is the wavelength of the Cu K_α radiation, and h, k, and l are the Miller indices for the diffraction peak. When multiple CdTe ZB peaks were present, the a_{Cubic} values were plotted as a function of the Nelson-Riley-Sinclair-Taylor (NRST) function [6],

$$NRST = \frac{1}{2}\left(\frac{\cos^2\theta}{\sin\theta} + \frac{\cos^2\theta}{\theta}\right).$$

Fitting a line to these points enabled a lattice constant of greater precision to be found by extrapolating to normal incidence at $NRST = 0$ [6]. Vegard's law, which applies to the CdTe-CdS alloy system [4], was then used to determine the mole fraction, x, for the cubic ZB phase:

$$x = \frac{a_{CdS_xTe_{1-x}(Cubic)} - a_{CdTe}}{a_{CdS} - a_{CdTe}}.$$

For cubic phases, $a_{CdTe} = 6.481$ Å and $a_{CdS} = 5.818$ Å.

Fig. 3 compares the x values obtained from EPMA and XRD for CdS_xTe_{1-x} alloy films sputtered in 100% Ar and subjected to a $CdCl_2$ HT. EPMA values measure the composition of the whole film. At higher S contents, EPMA values strongly exceed those obtained using the XRD analysis of the CdTe cubic ZB peaks, which measures the composition of the ZB phase only. The difference indicates that the film composition is within the CdS/CdTe miscibility gap for films of the higher two S contents. The miscibility gap was measured to lie between x = 0.058 and x = 0.97 [7] at temperatures of 415°C. Mole fraction values measured by XRD for the Te-rich cubic ZB phase in this study are consistent with the x = 0.058 observed by Jensen *et al.* [7]. Similar x values were observed for the evaporated films also (not shown).

To obtain information about grain orientation and size in these films, EBSD measurements were performed (Fig. 4) on CdS_xTe_{1-x} alloy films grown on Corning 7059/SnO₂:F/SnO₂/CdS:O substrates. One sputtered film and one evaporated film were measured after a $CdCl_2$ HT and a 0.5%-concentrated bromine-methanol etch for 2 s. Grains of both films are randomly oriented, indicated by the different colors of each grain. Pole figures (not shown) confirm this random orientation. The grain size of the sputtered film is 360 nm, whereas that of the evaporated film is

Table 2. Phases present in evaporated CdS_xTe_{1-x} films before and after a $CdCl_2$ HT.

CdS Content	As Deposited	After CdCl$_2$
Lower (x = 0.01 to 0.12 as deposited)	CdTe ZB (111)	CdTe ZB Minor CdS WZ
Higher (x = 0.12 to 0.23 as deposited)	CdTe ZB (111), shifted toward CdS WZ (100) at high S	CdTe ZB CdS WZ

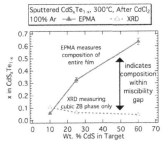

Fig. 3. Mole fraction (x) obtained from EPMA and XRD measurements for sputtered CdS_xTe_{1-x} films.

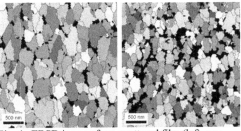

Fig. 4. EBSD images for a sputtered film (left; x = 0.059, grain size 360 nm) and an evaporated film (right; x = 0.28, grain size 240 nm). Black lines indicate grain boundaries; red lines indicate ∑3 twin boundaries. Both films were imaged after a $CdCl_2$ HT.

240 nm. This difference may be due to the greater adatom surface energy of the sputtering process. The $CdCl_2$ HT temperature for the sputtered film was also 10°C higher (400°C vs. 390°C), which may have contributed to increased recrystallization.

The BG of sputtered and evaporated CdS_xTe_{1-x} films was calculated by measuring the reflectance and transmittance, calculating the absorption coefficient (α) [9], and plotting $(\alpha h\nu)^2$ vs. hν, where hν is the photon energy. The evaporated films appear to have similar BG values (Fig. 5), indicating that these films have phase separated due to the miscibility gap. The sputtered films deposited in the 100% Ar ambient also appear to have values near the 1.5 eV BG of CdTe. An interesting phenomenon was observed for the films deposited in the 1% O_2/Ar ambient. As deposited, these films had BG values much higher than expected. In fact, the 60/40 wt.% CdS/CdTe film had a BG value higher than that of CdS itself. We suspect this behavior is similar to that observed for CdS:O by Wu et al. [2], in which CdS films were sputtered in O_2 partial pressure. In that study, amorphous CdS:O films with BGs of up to 3.1 eV were deposited. The increase in BG was attributed to nanocrystalline quantum size effect

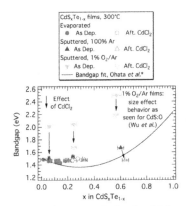

Fig. 5. Bandgap vs. mole fraction (x) for sputtered and evaporated CdS_xTe_{1-x} films. The black curve was determined by Ohata et al. [5] by fitting experimental data.

behavior. We believe a similar effect is occurring in CdS_xTe_{1-x} films deposited in O_2 partial pressure. After a $CdCl_2$ HT, the low-S film decreases to a BG near that of CdTe, whereas the higher-S films decrease in BG substantially, but not to the expected levels. This decrease is consistent with partial grain recrystallization and growth occurring as a result of the HT, although it seems to be incomplete.

CONCLUSIONS

CdS_xTe_{1-x} alloy films were deposited by RF magnetron sputtering and co-evaporation. As-deposited sputtered films grown in 100% Ar from targets containing 10/90 and 25/75 wt.% CdS/CdTe have a cubic CdTe ZB structure, whereas those grown from a target of 60/40 wt.% CdS/CdTe have a hexagonal CdS WZ structure. Films become less preferentially oriented as a result of a vapor $CdCl_2$ HT at 400°C for 5 min. Films sputtered in a 1% O_2/Ar ambient are amorphous as deposited, but show CdTe ZB, CdS WZ, and CdTe oxide phases after a $CdCl_2$ HT. Evaporated films primarily consist of the CdTe ZB phase as deposited, but the CdTe ZB (111) peak is shifted significantly toward the adjacent CdS WZ (100) peak for films of higher S content. These two peaks become distinct after a $CdCl_2$ HT, indicating phase separation has occurred. Both sputtered and evaporated films have randomly oriented grains after a $CdCl_2$ HT. The grain size of the sputtered film is larger than that of the evaporated film (360 vs. 240 nm). CdS_xTe_{1-x} alloy films sputtered in 1% O_2/Ar are amorphous as deposited and have a much higher bandgap than expected. This may be explained by nanocrystalline size effects seen previously [2] for CdS:O films.

Future work will include additional EBSD measurements to identify CdS_xTe_{1-x} phases and their intermixing both before and after a $CdCl_2$ HT. Auger electron spectroscopy and X-ray photoelectron spectroscopy will be used to examine the CdTe oxides that result from the $CdCl_2$ HT of films grown in 1% O_2/Ar. In future PV device work, we plan to replicate the existing superstrate device structure using directly deposited CdS_xTe_{1-x} layers in PV devices. We also expect to design and deposit new superstrate and substrate PV device structures using these directly deposited layers.

ACKNOWLEDGEMENTS

This work was supported under DOE Contract No. DE-AC36-08-G028308 to NREL.

REFERENCES

1. R.G. Dhere, Y. Zhang, M.J. Romero, S.E. Asher, M. Young, B. To, R. Noufi, and T.A. Gessert, *Proc. of the 33rd Photovoltaic Specialists Conference* (IEEE, San Diego, CA, 2008).
2. X. Wu, Y. Yan, R.G. Dhere, Y. Zhang, J. Zhou, C. Perkins, and B. To, *phys. stat. sol. (c)* **1**, 1062-1066 (2004).
3. R.G. Dhere, Ph.D. Thesis, University of Colorado, 1997.
4. K. Ohata, J. Saraie, and T. Tanaka, *Japan. J. Appl. Phys.* **12**, 1198-1204 (1973).
5. K. Ohata, J. Saraie, and T. Tanaka, *Japan. J. Appl. Phys.* **12**, 1641-1642 (1973).
6. D.G. Jensen, Ph.D. Thesis, Stanford University, 1997.
7. D.G. Jensen, B.E. McCandless, and R.W. Birkmire, *Proc. of the 25th Photovoltaic Specialists Conference* (IEEE, Washington, D.C., 1996).
8. B.E. McCandless, G.M. Hanket, D.G. Jensen, and R.W. Birkmire, *J. Vac. Sci. Technol. A* **20**, 1462-1467 (2002).
9. J.I. Pankove, *Optical Processes in Semiconductors* (Dover Publications, Inc., New York, NY, 1971).

Mater. Res. Soc. Symp. Proc. Vol. 1324 © 2011 Materials Research Society
DOI: 10.1557/opl.2011.963

Influence of Annealing in H_2 Atmosphere on the Electrical Properties of Thin Film CdS

Natalia Maticiuc[1, 2], Jaan Hiie[1], Tamara Potlog[2], Vello Valdna[1], Aleksei Gavrilov[1]
[1]Department of Materials Science, Tallinn University of Technology, 5 Ehitajate tee, Tallinn, 19086, Estonia.
[2]Department of Applied Physics and Informatics, Moldova State University, 60 A. Mateevici street, Chisinau, MD 2009, Moldova.

ABSTRACT

Chemical bath-deposited (CBD) thin film CdS has been widely used as a buffer and n-type window layer in CdS/CIGS and CdS/CdTe thin film solar cells. Annealing of CBD CdS assigns to the layers required concentration and mobility of electrons, crystallinity, structural stability and perfect Ohmic front contact in TCO/CdS interface. But always annealing reduces band gap (E_g) of solution-deposited CdS and lowers current density of the CdS/CdTe PV device due to optical absorption within the CdS layer.

We have studied systematically dynamics of changes in CBD CdS/glass thin film structural, optical and electrical properties in annealing process in H_2 ambient at normal pressure in pre-heated ceramic tubular furnace. Here we'll present electrical, characterization results of annealed CBD CdS/glass thin films, 300 nm thick. The films were deposited with thiourea from ammoniacal 1 mM dilute solution of $CdSO_4$ and 0.001 at. % of NH_4Cl relative to Cd for Cl doping.

We found high concentration of electrons 1-4 E19 cm^{-3} in the layers annealed at 200 - 450 °C, while for 200 °C the long time of annealing over 60 min is needed, but for high temperature region 350 - 450 °C only for short 10 min annealing this concentration region of electrons was achieved. In the high temperature region rapid decrease of electron concentration and conductivity will go on with increasing annealing temperature and time. Mobility of electrons will decrease from 9 to 5 $cm^2/V \cdot s$ in the annealing region 200-350 °C, which is probably connected with disordering of lattice. On the basis of acquired results we propose an hypothesis about substitutional incorporation of OH group on S site in CdS lattice in deposition process and that $(OH)_s$ complex defect acts as a donor defect like Cl and we believe that the both defects are responsible for changes of thin film CdS electrical, optical and structural properties in the annealing process. Thermal annealing in hydrogen atmosphere is a convenient and appropriate method for precise control of CdS thin film electrical properties, and also for creation of n/n^+ CdS window layers in the substrate configuration of a solar cell.

INTRODUCTION

CdS thin films are used as n-type window layers in CdTe [1] and CIGS [2] based solar cells due to wide band gap, high transparency and reasonable mobility of electrons. The performance of solar cells that use chemically deposited CdS are superior to those using evaporated CdS films [3]. This is primarily due to a shift of the absorption edge of the film to shorter wavelengths, in comparison with PVD (physical vapor deposited) CdS [4] – an advantage in the role of as deposited CBD CdS thin films as window materials in CIGS based substrate type heterojunction solar cells [2]. However, CBD CdS performs well also in annealed superstrate configuration FTO/CdS/CdTe solar cells [5], though in annealing process the band gap decreases from about 2.45 eV to 2.2 eV resulting in lower photovoltaic response of solar cell [1] due to increased absorption in window layer. The formation of heterojunction depends on concentration and conductivity of charge carriers and on the quality of interface between n- and p- type components of the junction. Several authors have found that annealing of CBD CdS thin films at low temperatures $120 - 200\,^{\circ}C$ improves solar cell parameters [6, 7]. These circumstances impose for further more detailed investigations of CdS annealing processes.

In [8] we described results of systematic investigations of changes in structural and optical properties of CBD CdS resulting from post-deposition thermal treatment in hydrogen ambient and trying to understand the solution chemistry and how the properties of CdS thin films could be tailored for control of the final solar cell characteristics. In this work we report the influence of the annealing on electrical properties of CdS thin films deposited by CBD method.

EXPERIMENT

CdS thin films were deposited on glass substrates in a hermetically closed glass jar in water solution of $CdSO_4$ (1 mM), thiourea (10 mM), NH_4OH (0.2 M), $(NH_4)_2SO_4$ (0.03 M) and relative to Cd 0.001 at % NH_4Cl at $80\,^{\circ}C$. All samples were deposited three times. The CBD CdS layers were washed and dried at $120\,^{\circ}C$ in vacuum and these layers serve as initial samples (as deposited in the text) for annealing in a preheated tubular furnace in a quartz process tube under 1 atm of H_2 in the temperature range of $200 - 450\,^{\circ}C$ for 3 - 120 minutes. The heat treatment in H_2 provides reducing atmosphere for deoxidation of CdS layers, as well its normal pressure inhibits evaporation of chlorine dopant from CdS. It has to be emphasized that each experimental point of annealing has been made with new as deposited samples. The CdS thin films were characterized by x-ray diffraction (XRD) in the ω - $2v$ configuration (Cu-kα) (Rigaku X-ray - LAST IV), scanning electron microscopy (SEM) (Zeiss EVO MA-15), energy-dispersive x-ray spectroscopy (EDS) (Link Analytical AN 10000), optical reflection and transmission on Jasco-V-670 type spectrophotometer. Sheet resistance and Hall effect were measured by Van der Paw four probe method at room temperature. In-plane resistivity and charge carrier concentrations were calculated for the thicknesses (280 - 430 nm) estimated from transmission and reflection data.

RESULTS

At all annealing temperatures a rapid decrease of resistivity of annealed samples by 4-5 orders of magnitude and acceleration of the decay rate with increasing of annealing temperature has been observed, as shown in Figure 1. The abrupt decay in 10 minutes is followed by slow changes of resistivity at longer annealing times and could be divided into three sub-regions of moderate (200-250 $^{\circ}C$), intermediate (250-350 $^{\circ}C$) and high (350-450 $^{\circ}C$) temperatures. In the low temperature area at 200 $^{\circ}C$ the lowest resistivity of 0.02 $\Omega\cdot cm$ is achieved and stabilized at 60 min. annealing time. The stabilization time is shortening at higher temperatures. In intermediate region, after the rapid decrease, the resistivity will increase in growing rate with increasing of annealing temperature.

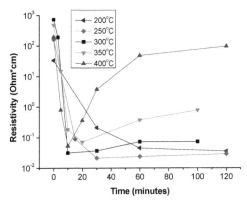

Figure 1. Resistivity of CdS layers versus annealing temperature and time.

Hall mobility was measured for samples with sheet resistance lower than 3 kΩ/□, see Table 1. Lowest values of mobility lie in the intermediate region of annealing temperature. Concentration of electrons is very high, > 10^{19} cm^{-3}, for low and intermediate regions of temperatures.

Table 1. Electrical properties of annealed CdS thin films. (The first number in the name of sample corresponds to the annealing temperature, $^{\circ}$C – and annealing time in minutes)

Sample ($^{\circ}$C-min)	Thick-ness (nm)	Sheet Resistance (k Ω /□)	Electron Mobility (cm^2/V·s)	Resistivity (×10^{-2}, Ω·cm)	Electron Concentration (×10^{19}, cm^{-3})
200 - 60	430	1.06 ± 0.01	9.3 ± 0.4	4.57 ± 0.01	1.5 ± 0.1
200 - 120	420	0.839 ± 0.003	9.0 ± 1.5	3.52 ± 0.02	2.0 ± 0.3
250 - 30	370	0.582 ± 0.004	9.5 ± 0.6	2.15 ± 0.02	3.1 ± 0.2
250 - 60	370	0.648 ± 0.002	8.0 ± 0.7	2.40 ± 0.01	3.3 ± 0.3
250 - 120	370	0.776 ± 0.007	4.0 ± 1.5	2.88 ± 0.02	4.1 ± 1.5
300 - 10	280	1.13 ± 0.01	5.4 ± 0.8	3.17 ± 0.01	3.7 ± 0.6
300 - 60	280	2.61 ± 0.01	3.5 ± 0.2	7.31 ± 0.03	2.46 ± 0.15
300 - 100	290	2.51 ± 0.02	3.8 ± 0.2	7.27 ± 0.05	2.28 ± 0.15
350 - 20	370	1.95 ± 0.02	4.5 ± 0.2	7.22 ± 0.15	1.9 ± 0.1
400 - 10	390	1.36 ± 0.01	4.5 ± 0.2	5.03 ± 0.02	2.76 ± 0.15

DISCUSSION

The most striking result is creation of high concentration of electrons $4.1 \cdot 10^{19}$ cm^{-3} at moderate annealing temperatures 200 – 250 °C, see Table 1. An abrupt decrease of resistivity and increase of electron concentration in CBD CdS thin films annealed in hydrogen or in vacuum at the same moderate temperatures has also been observed by [9, 10]. It should be mentioned that this fact may be a possible reason for formation of reverse diode in the n$^+$-CdS/i-ZnO/ZnO:Al window region of CIGS type substrate configuration solar cells like in the case for n$^+$-CdO/n-CdS/ITO interface described by Durose [11].

The high concentration of electrons corresponds to the high density of shallow donors in CdS lattice and the resistivity of such degenerated semiconductor layers does not depend on temperature (Figure 2). The high conductivity cannot be explained by creation of conductive phase of CdO from the residual oxide components, because reduction of CdO begins intensively from 200 °C [12]. Also it can not explained by creation of sulfur vacancies because long H$_2$ annealing at high temperature region results in high resistivity film, which has very low intensity of photoluminescence below band gap energy [13] indicating to the absence of impurities and complexes with cadmium vacancies, as potential acceptors.

Temperature dependence of conductivity for CdS thin film annealed at 400 °C has an activation energy of 0.2 eV over room temperature, which indicates to the presence of deep donors (Figure 2).

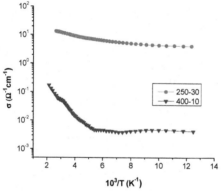

Figure 2. Conductivity of CdS thin films annealed at 250 and 400 °C versus reverse temperature of measurement, 10^3/T (K^{-1}).

Increase in annealing temperature accelerates the fall of resistivity. As a working hypothesis, incorporation of cadmium hydroxychloride into CdS lattice on sulfur site in deposition process from basic ammonia solution, is assumed. Presence of hydroxide group in as deposited CBD CdS thin films has been spectrometrically (FTIR) confirmed by Kylner [14]. Sulfur, chlorine and hydroxide group are spatially compatible, their atomic sizes are about 100 pm, for oxygen 60 pm (compare densities of cubic modifications of CdS 4.82 g/cm^3 and CdO 8.2 g/cm^3), and for hydrogen 30 pm. The hydroxide and chloride ions on sulfur site form electrically neutral defect complexes $V_{Cd}2Cl_s$ and $V_{Cd}2(OH)_s$. Increase in annealing temperature accelerates sintering of as deposited polycrystalline CdS and retention of larger quantities of chloride and hydroxide impurities, which would otherwise evaporate and decompose. Upon decomposition of two hydroxide groups in CdS lattice one water molecule will be released and one sulfur vacancy created. Annihilation of sulfur and cadmium vacancies creates two donor centers of Cl_s^+ and $(OH)_s^+$. These substitutional impurities on sulfur site behave as shallow donors contributing at room temperature to the decrease of resistivity. The concentration of chloride and hydroxide impurities, saved in as deposited CdS, depends on post-deposition temperature of vacuum drying. At high temperature region of annealing, due to destruction of hydroxide group and out diffusion of chlorine and removal by evaporation, the concentration of electrons decreases.

Incorporation of cadmium chloride could be sketched with equations 1 – 5:

$$CdCl_2 \text{ (solution)} \rightarrow Cd_{Cd}{}^x + (V_{Cd}2Cl_s) \tag{1}$$

$$(1+n)(V_{Cd}2Cl_s)^x \rightarrow V_{Cd}{}^{2-} + 2ne^- + 2(1+n)Cl_s^+ + nS_{2\,(g)}\uparrow \tag{2}$$

$$V_{Cd}{}^{2-} + Cl_s^+ \rightarrow (V_{Cd}Cl_s)^- \tag{3}$$

$$CdCl_{2(g)} + Cd_{(g)} = 2Cd_{Cd}{}^x + 2Cl_s^+ + 2e^- \tag{4}$$

$$V_{Cd}{}^{2-} + 2Cl_s^+ + Cd_{Cd}{}^x \rightarrow CdCl_{2(g)}\uparrow \tag{5}$$

For all annealing temperatures until 400 °C a technologically important stability region could be envisaged, for example at 400 °C it will be 60 minutes or longer. All annealed samples have been electrically stable over a year.

Thermal annealing in hydrogen atmosphere is a convenient and appropriate method for precise control of CdS thin film electrical properties, and also for creation of n/n$^+$ CdS window layers in the substrate configuration of a solar cell.

CONCLUSIONS

Thermal annealing in hydrogen atmosphere is a convenient and appropriate method for precise control of CdS thin film electrical properties, and also for creation of n/n$^+$ CdS window layers in the substrate configuration of a solar cell.

ACKNOWLEDGMENTS

This work has been supported by EU 7th FP project FLEXSOLCELL GA-2008-230861, by Estonian Science Foundation grants 7241 and 7608, and by Estonian National Target Financing Nos SF0140092s08, and SF0140099s08.

REFERENCES

1. X. Wu, *Solar Energy* **77**, 803–814 (2004).
2. I. Repins, M. Contreras, M. Romero, Y. Yan, W. Metzger, J. Li, S. Johnston, B. Egaas, C. DeHart, J. Scharf, B. E. McCandless and R. Noufi, *PVSC '08 33rd IEEE Proc.*, 1-6 (2008).
3. M. T. S. Nair, P. K. Nair, R. A. Zingaro and E. A. Meyers, *J.Appl.Phys.*, **75(3)**, 1557-1564 (1994).
4. D. Abou-Ras, G. Kostorz, A. Romeo, D. Rudmann and A.N. Tiwari, *Thin Solid Films*, **480–481**, 118–123 (2005).
5. R. G. Dhere, J. N. Duenow, A. Duda, S. Glynn, J. Li, W. K. Metzger, H. Moutinho and T. A. Gessert, *Mater. Res. Soc. Symp. Proc.*, **1165**, 1165-M02-08 (2009).
6. T. Sakurai, N. Ishida, S. Ishizuka, M. M. Islam, A. Kasai, K. Matsubara, K. Sakurai, A. Yamada, K. Akimoto and S. Niki, *Thin Solid Films*, **516**, 7036–7040 (2008).
7. Y. D. Chung, D. H. Cho, N. M. Park, K. S. Lee and J. Kim, *Current Applied Physics*, **11**, S65-S67 (2011).
8. N. Maticiuc, T. Potlog, J. Hiie, V. Mikli, N. Põldme, T. Raadik, V. Valdna, A. Mere, A. Gavrilov, F. Quinci, V. Lughi and V. Sergo, *Moldavian Journal of the Physical Sciences*, **9(3-4)**, 275-279 (2010).
9. J. Hiie, K. Muska, V. Valdna, V. Mikli, A. Taklaja and A. Gavrilov, *Thin Solid Films*, **516**, 7007-7012 (2008).
10. E. Vasco, E. Puron and O. de Melo, *Material Letters*, **25**, 205-207 (1995).
11. M. K. Al Turkestani and K. Durose, *Sol. Energy Mat. & Solar Cells*, **95**, 491-496 (2011).
12. HSC Chemistry 4.0, Chemical Reaction and Equilibrium Software with extensive Thermochemical Database (1999).
13. J. Hiie, F. Quinci, V. Lughi, V. Sergo, V. Valdna, V. Mikli, E. Kärber and T. Raadik, *Mater. Res. Soc. Symp. Proc.*, **1165**, 1165-M08-17 (2009).
14. A. Kylner, A. Rockett and L. Stolt, *Sol.Stat. Phenom.*, **51-52**, 533-540 (1996).

CIGS/CIS

Mater. Res. Soc. Symp. Proc. Vol. 1324 © 2011 Materials Research Society
DOI: 10.1557/opl.2011.1152

Properties of CuIn₁₋ₓGaₓSe₂ films prepared by the rapid thermal annealing of spray-deposited CuIn₁₋ₓGaₓS₂ and Se

Laura E. Slaymaker[1], Nathan M. Hoffman[1], Matthew A. Ingersoll[1], Matthew R. Jensen[1], Jiří Olejníček[1], Christopher L. Exstrom[1], Scott A. Darveau[1], Rodney J. Soukup[2], Natale J. Ianno[2], Amitabha Sarkar[2], and Štěpán Kment[2]

[1] Department of Chemistry, University of Nebraska at Kearney, Kearney, NE 68849, U.S.A.

[2] Department of Electrical Engineering, University of Nebraska-Lincoln, Lincoln, NE 68588-0511, U.S.A.

ABSTRACT

Many reported CuIn₁₋ₓGaₓSe₂ (CIGS) thin films for high-efficiency solar cells have been prepared via a two-stage process that consists of a high-vacuum film deposition step followed by selenization with excess H₂Se gas or Se vapor. Removing toxic gas and high-vacuum requirements from this process would greatly simplify it and make it less hazardous. We report the formation of CuIn₁₋ₓGaₓSe₂ (x = 0, 0.25, 0.50, 0.75, 1.0) thin films achieved by rapid thermal annealing of spray-deposited CuIn₁₋ₓGaₓS₂ and Se in the absence of an additional selenium source. To prepare the Se layer, commercial Se powder was dissolved by refluxing in ethylenediamine/2,2-dimethylimidizolidine. After cooling to room temperature, this mixture was combined with 2-propanol and the resulting colloidal Se suspension was sprayed by airbrush onto a heated glass substrate. The resulting film was coated with nanocrystalline CuIn₁₋ₓGaₓS₂ via spray deposition of a toluene-based "nanoink" suspension. The two-layer sample was annealed at 550 °C in an argon atmosphere for 60 minutes to form the final CIGS product. Scanning electron microscopy images reveal that film grains are 200-300 nm in diameter and comparable to sizes of the reactant CuIn₁₋ₓGaₓS₂ nanoparticles. XRD patterns are consistent with the chalcopyrite unit cell and calculated lattice parameters and A₁ phonon frequencies change nearly linearly between those for CuInSe₂ and CuGaSe₂.

INTRODUCTION

Chalcopyrite CuIn₁₋ₓGaₓSe₂ (CIGS) thin films for use as absorbers in high-efficiency photovoltaic devices have been prepared via a two-stage process in which a thin film of sputtered or thermally evaporated CuIn₁₋ₓGaₓ is reacted with H₂Se or Se vapor at high temperature [1-3]. Efforts to lower productions costs by eliminating high-vacuum processing steps have stimulated research in solvothermal methods of CIGS preparation. While chemically pure nanocrystalline CIGS may be prepared in this manner [4-6], the removal of solvent from films cast from nanoparticle suspensions (or "nanoinks") by heating results in films with voids and small grains that lead to low energy conversion efficiencies [7].

Guo et. al. [7] have demonstrated that a dense thin film of chalcopyrite-phase CuInSe₁.₈S₀.₂ may be prepared by reaction of H₂Se gas with a film cast from a CuInS₂ nanoink suspension. Replacement of the S²⁻ ions with the larger Se²⁻ ions reduces void formation and device efficiencies as high as 5.55% have been reported [7]. In selenization reactions such as this, atmospheres of excess H₂Se or Se vapor are typically maintained over the CuIn₁₋ₓGaₓ alloy films in order to generate CIGS films of uniform composition. In the analogous preparation of CuIn₁₋ₓAlₓSe₂ (CIAS), a ternary chalcopyrite with a more strained crystal lattice, the presence of argon gas in the selenization atmosphere can sufficiently inhibit Se diffusion from the heated product film, enabling the formation of uniform CIAS [8,9] and overcoming the problem of ordered defect compound formation within the film [9]. This observation has led our group to explore a new approach for preparing CIGS thin films that involves the deposition of complexed selenium nanoparticles onto a substrate followed by spraying of $CuIn_{1-x}Ga_xS_2$ (x = 0, 0.25, 0.5, 0.75, 1) particles with an airbrush. Rapid thermal annealing of these layered nanoparticles in an argon atmosphere results in displacement of sulfur by selenium forming chalcopyrite $CuIn_{1-x}Ga_xSe_2$ without the need for additional Se in the reaction system.

EXPERIMENTAL DETAILS

All starting chemicals were purchased commercially (Aldrich, Fisher, Alfa Aesar reagent grade) and used as received. Reactions to prepare $CuIn_{1-x}Ga_xS_2$ (x = 0, 0.25, 0.5, 0.75, 1) nanoparticles and the Se complex solution were conducted in an argon atmosphere. Nanoink suspensions of $CuIn_{1-x}Ga_xS_2$ in toluene were prepared as described in the literature [7]. A colloidal Se spray solution was prepared by complexing Se with a condensation product of ethylenediamine (EDA) and 2,2-dimethylimidizolidine. A mixture of 2,2-dimethylimidizolidine (10.0 mL), EDA (9.1 mL), and toluene (5.7 mL) was refluxed at 85 °C for 2 hours. The resulting condensation product (20.0 mL) was collected in a Dean-Stark trap and a portion (6.0 mL) was combined with EDA (4.1 mL) and Se powder (200 mesh, 0.3959 g). This mixture was refluxed with continuous stirring for one hour. After cooling to room temperature, a 2-mL portion was combined with 40 mL of 2-propanol that had been heated to 45 °C for 30 minutes. Within one minute, the resulting brown mixture turned a brilliant red color indicative of colloidal Se formation.

Films of Se and $CuIn_{1-x}Ga_xS_2$ were squentially sprayed onto a plain or Mo-coated sodalime glass microscope slide using a Paasche VLS airbrush with D500 model air compressor. Initially, the target slide was placed on a second microscope slide on an aluminum foil-covered hot plate that was heated to 120 °C. The colloidal Se solution was then sprayed on the heated microscope slide at a rate slow enough to allow the solvent to evaporate immediately on contact with the slide. The initial Se film was red in color, gradually turning dark gray. The slide was removed from the hot plate to cool to room temperature on the bench top when spraying was complete. The $CuIn_{1-x}Ga_xS_2$ nanoink suspension was sprayed by airbrush onto the Se-coated slide in a vertical position until a 1:1 mole ratio was achieved as determined by weighing the sample after solvent evaporation. In some samples on Mo-coated slides, the $CuIn_{1-x}Ga_xS_2$ and Se were spray-deposited in reverse order.

To prepare a $CuIn_{1-x}Ga_xSe_2$ (x = 0, 0.25, 0.5, 0.75, or 1) film, the $CuIn_{1-x}Ga_xS_2$/Se coated slide was placed in a graphite container. This was loaded into a quartz tube that was evacuated using a rotary mechanical pump to a base pressure of less than 1 Pascal and then filled with 1 atm of argon (99.999%). A thermocouple was inserted into the body of the graphite container and using a Quad Ellipse Chamber Heater connected to a Model 915 power supply temperature controller from Research, Inc., the sample was heated for 60 minutes at 550°C. After cooling to room temperature, the sample was removed from the quartz tube and graphite container.

All Se, $CuIn_{1-x}Ga_xS_2$ and $CuIn_{1-x}Ga_xSe_2$ materials were characterized by Raman spectroscopy (Horiba/Jobin Yvon LabRAM HR800 with He-Ne laser), X-ray diffraction (XRD, Bruker AXS D8 Discover), and scanning electron microscopy (SEM, Hitachi S4700).

DISCUSSION

We report the successful reaction between layered nanocrystalline $CuIn_{1-x}Ga_xS_2$ (x = 0, 0.25, 0.5, 0.75, 1) and Se materials to form the respective $CuIn_{1-x}Ga_xSe_2$ thin films. Both of the precursor $CuIn_{1-x}Ga_xS_2$ and Se materials were prepared via solution-based methods that have potential to be scaled up using standard chemical engineering techniques. The rapid thermal annealing of these layered precursors was accomplished in an argon atmosphere without the need for an external supply of H_2Se or Se.

$CuIn_{1-x}Ga_xS_2$ characterization

Scanning electron microscopy (SEM) images (Figure 1) of the $CuInS_2$ product reveal that the particles are large in diameter (200-500 nm) and have generally cubic morphologies. The X-ray diffraction pattern confirms the chalcopyrite crystal structure. The Raman spectrum shows a 295-cm^{-1} signal that is characteristic of the $CuInS_2$ A_1 phonon frequency while the signal at 305 cm^{-1} indicates the presence of Cu-poor regions in the material [10]. This did not appear to affect the quality of the product upon reaction with selenium.

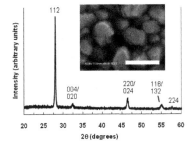 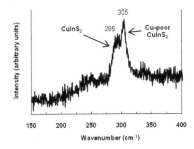

Figure 1. XRD pattern (left), SEM image (inset, white scale bar = 1.00 μm), and Raman spectrum (right) of solvothermally prepared $CuInS_2$.

When preparing $CuGaS_2$, 0.50 mmol of $Ga(NO_3)_3$ was used in place of $InCl_3$ in the reaction. The A_1 phonon frequency in the isolated product is shifted to 310 cm^{-1} as expected from the smaller crystal unit cell that results from substituting In^{3+} with Ga^{3+} [11]. Investigations of Ga/(In+Ga) ratio control are underway. To achieve compositions of $CuIn_{1-x}Ga_xS_2$ (x = 0.25, 0.50, 0.75), mixtures of $Ga(NO_3)_3$ and $InCl_3$ were used in the reaction such that their mole ratio matched that of the nominal composition and the total amount used equaled 0.50 mmol. All precursor materials described in this section were found to be suitable for reaction with Se.

Colloidal Se formation and characterization

Recently, it has been reported that commercial Se powder may be solubilized in refluxing ethylenediamine (EDA) and precipitated as 1-D nanoparticles upon injection of this solution into water or any non-amine organic solvent [12]. It is theorized that the solubilization is due to the formation of $Se(EDA)_n$ complexes [13]. In a precipitating solvent, the EDA dissociates and Se nanoparticle nucleation and growth occur. In our laboratory, we have discovered that precipitated Se particles from this complex aggregate too rapidly to produce sprayable Se films suitable for contact and reaction with $CuIn_{1-x}Ga_xS_2$ nanoparticles. The reaction of commercial Se powder with the 2,2-dimethylimidizolidine/EDA condensation product appears to produce a mixture of soluble organoselenium compounds that may include selones, diacylselenides and/or diacyldiselenides according to ^{1}H NMR evidence. Dilution in alcohol solvents hydrolyzes the organic condensation product and releases Se that forms colloidal particles. The degree of particle aggregation varies with the nature of the dilution solvent. The use of heated (45 °C) 2-propanol has resulted in Se colloidal suspensions that produce the most uniform films via airbrush spraying.

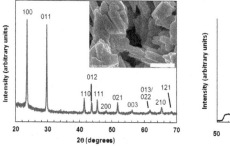 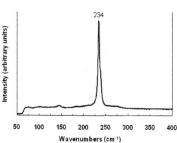

Figure 2. XRD pattern (left), SEM image (inset, white scale bar = 1.00 μm), and Raman spectrum (right) of a typical Se film sprayed from colloidal suspension.

Uniform gray trigonal crystalline films, as evidenced by Raman spectroscopy and XRD (Figure 2), result from the spraying of the above-described Se colloidal suspension on sodalime glass substrates heated to 120 °C. SEM images (Figure 2 inset) reveal nano/microrod morphologies with narrow diameters of 200-300 nm and aspect ratios ranging from 2 and 5.

$CuIn_{1-x}Ga_xSe_2$ film formation and characterization

Upon heating the dual-layered $Se/CuIn_{1-x}Ga_xS_2$ precursor films at 550 °C for one hour in a 1-atm argon atmosphere, the Se diffuses through the $CuIn_{1-x}Ga_xS_2$ layer, displacing the sulfur to form the final CIGS product. XRD patterns of the product $CuIn_{1-x}Ga_xSe_2$ films (Figure 3) are consistent with the chalcopyrite unit cell and show diffractions from the (112), (220/024), and (116/132) planes. All peak positions shift toward higher 2θ values as the Ga/(In+Ga) ratio increases. Figure 3 also shows this in greater detail for the (112) plane spacing. Calculated lattice parameters change nearly linearly between those for $CuInSe_2$ (a = 5.791 Å, c = 11.585 Å) and $CuGaSe_2$ (a = 5.637 Å, c = 10.992 Å).

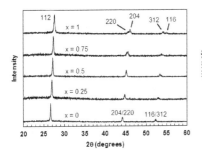 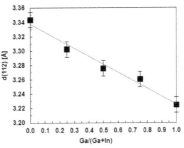

Figure 3. (Left) XRD patterns for product $CuIn_{1-x}Ga_xSe_2$ (x = 0, 0.25, 0.50, 0.75, 1) films and (right) (112) lattice plane spacing (Å) as a function of Ga/(Ga+In).

The Raman spectra (Figure 4) show a similarly linear shift in A_1 phonon frequency between that for $CuInSe_2$ (171 cm^{-1}) and $CuGaSe_2$ (184 cm^{-1}). No residual $CuIn_{1-x}Ga_xS_2$, $Cu_{2-x}Se$, or $Cu_{2-x}S$ phonon signals are present and these XRD and Raman data indicate that all sulfur may have been displaced from the films. SEM images (Figure 4) of $CuIn_{1-x}Ga_xSe_2$ films show grains that are 200-300 nm in diameter and comparable in size to the $CuIn_{1-x}Ga_xS_2$ precursor nanoparticles.

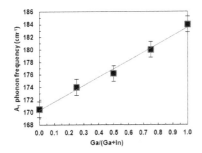 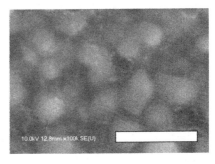

Figure 4. (Left) A_1 phonon frequencies for product $CuIn_{1-x}Ga_xSe_2$ films as a function of Ga/(Ga+In); (right) SEM image of product $CuInSe_2$ film (white scale bar = 1.00 μm)

When excess $CuInS_2$ was used, Cu-rich CuAu-ordered $Cu_{1.5}InSe_2$ phase resulted. The additional copper shifted the $CuInSe_2$ A_1 phonon frequency that typically appears at about 173 cm^{-1} to about 175 cm^{-1} and broadened the peak with a shoulder at 183 cm^{-1}, a peak characteristic of the copper rich $Cu_{1.5}InSe_2$ phase [14].

In working toward the fabrication of a PV device prototype, the layering of nanocrystalline $CuInS_2$ and colloidal Se on Mo-coated sodalime glass substrates was attempted. When Mo/Se/$CuInSe_2$ layered slides were annealed as described above, the final product was mostly $MoSe_2$. Changing the layering order to Mo/$CuInS_2$/Se seemed to slow the formation of $MoSe_2$

and some $CuInSe_2$ material was created. It was found that a precise 2:1 mole ratio of $Se:CuInS_2$ was necessary or unwanted phases of $CuInSe_2$ and binary $Cu_{2-x}Se$ phases would result.

CONCLUSIONS

We report the formation and characterization of $CuIn_{1-x}Ga_xSe_2$ (x = 0, 0.25, 0.5, 0.75, 1) thin films prepared by the rapid thermal annealing of layered $CuIn_{1-x}Ga_xS_2$ nanoparticles and colloidal Se in an argon atmosphere. Sulfur displacement by selenium results in the formation of chalcopyrite $CuIn_{1-x}Ga_xSe_2$ without the need for additional Se in the reaction system. Raman spectroscopy and XRD evidence support the chalcopyrite structure of the film and at all nominal stoichiometries, the Ga/(In+Ga) ratio is maintained during the reaction.

ACKNOWLEDGMENTS

Work supported by the U.S. Department of Energy (Grant DE-FG36-08GO88007) and the Nebraska Center for Energy Sciences Research.

REFERENCES

1. V. Alberts, J. H. Schön, M. J. Whitcomb, E. Bucher, U. Rühle, and H. W. Schock, *J. Phys. D: Appl. Phys.* **31**, 2869-2876 (1998).
2. R. Caballero and C. Guillén, *Thin Solid Films* **431-432**, 200-204 (2003).
3. V. F. Gremenok, E. P. Zaretskaya, V. B. Zalesski, K. Bente, W. Schmitz, *Int. Sci. J. Alt. Energy Ecol.* **5**, 39-43 (2004).
4. J. Tang, S. Hinds, S. O. Kelley, and E. H. Sargent, *Chem. Mater.* **20**, 6906-6910 (2008).
5. M. G. Panthani, V. Akhavan, B. Goodfellow, J. P. Schmidtke, L. Dunn, A. Dodabalapur, P. F. Barbara, and B. A. Korgel, *J. Amer. Chem. Soc.* **130**, 16770-16777 (2008).
6. J. Olejníček, C. A. Kamler, A. Mirasano, A. L. Martinez-Skinner, M. A. Ingersoll, C. L. Exstrom, S. A. Darveau, J. L. Huguenin-Love, M. Diaz, N. J. Ianno, and R. J. Soukup, *Sol. Energy Mater. Sol. Cells* **94**, 8-11 (2010).
7. Q. Guo, G. M. Ford, H. W. Hillhouse, and R. Agrawal, *Nano Lett.* **9**, 3060-3065 (2009).
8. J. Lopez-García, C. Guillén, *Thin Solid Films* **517**, 2240-2243 (2009).
9. J. Olejníček, C. A. Kamler, S. A. Darveau, C. L. Exstrom, L. E. Slaymaker, A. R. Vandeventer, N. J. Ianno, and R. J. Soukup, *Thin Solid Films* **519**, 5329-5334 (2011).
10. J. Álvarez-García, J. Marcos-Ruzáfa, A. Pérez-Rodríguez, A. Romano-Rodríguez, J. R. Morante, and R. Scheer, *Thin Solid Films* **361-362**, 208-212 (2000).
11. H. Matsushita, S. Endo, and T. Irie, *Jpn. J. Appl. Phys.* **31**, 18-22 (1992).
12. Z. Yang, S. Cingarapu, and K. J. Klabunde, *Chem. Lett.* **38**, 252-253 (2009).
13. J. Lu. Y. Xie, F. Xu, and L. Zhu, *J. Mater. Chem.* **12**, 2755-2761 (2002).
14. V. Izquierdo-Roca, X. Fontane, L. C. Barrio, J. Alvarez-Garcia, A. Perez-Rodriguez, J. R. Morante, W. Witte, and R. Klenk, *Mater. Res. Soc. Symp. Proc.* **1165**, M05-16 (2009).

Mater. Res. Soc. Symp. Proc. Vol. 1324 © 2011 Materials Research Society
DOI: 10.1557/opl.2011.964

I-III-VI₂ (Copper Chalcopyrite-based) Thin Films for Photoelectrochemical Water-Splitting Tandem-Hybrid Photocathode

Jess M. Kaneshiro[1], Alexander Deangelis[1], Xi Song[1], Nicolas Gaillard[1], Eric L. Miller[2]

[1]HNEI, U. of Hawai`i at Manoa, 1680 East-West Rd. POST109, Honolulu, HI 96822

[2]U.S. Department of Energy, EE-2H 1000 Independence Ave. SW, Washington, D.C. 20585

ABSTRACT

This presentation will investigate various parameters regarding the use of I-III-VI₂ Copper Chalcopyrite-based materials for use in tandem-hybrid photocathodes capable of splitting water into hydrogen and oxygen gases in an acidic electrolyte. Constituent parts (fabricated at HNEI) of a proposed monolithically integrated hybrid photovoltaic/photoelectrochemical (PV/PEC) device were characterized separately and combined theoretically using electronic and optical models to simulate tandem operation to first indicate feasibility of matching existing materials. Robust CGSe₂ photocathodes were focused on for the PEC cells and CIGSe₂ and CISe₂ devices were evaluated for the PV cells. Simulation suggested the hybrid PV/PEC system could pass enough light to produce up to 15.87mA/cm², validating the feasibility and warranting the fabrication of stacked PV/PEC devices.

INTRODUCTION

Materials in the I-III-VI₂ ternary chalcopyrite class (most popularly CuInGaSe₂) are highly suitable for use in solar cells because of very favorable optical and electronic properties as well as bandgap-tuneability based on alloy composition [1]. Tandem photovoltaic (PV) cells utilizing these materials have not yet achieved their theoretical performances, partially due to difficulty in utilizing the full potential of wide-bandgap top cells, mitigation of the shadowing of the bottom cells, and lattice matching between the many layers being incorporated [2,3,4,5]. Wide-bandgap thin films like CGSe₂ have been found to be very efficient solar energy conversion devices in a PEC setup (shown in Figure 1A), where the solar energy is used to directly split water into oxygen and hydrogen gases; the latter of which being a useful, convenient, and storable form of energy [6]. However, photoelectrochemical water-splitting with current materials require an applied voltage bias that can be very elegantly fulfilled by an underlying PV device to create a tandem-hybrid photocathode [7]. Two distinct advantages a PV/PEC tandem water-splitting cell would have over a tandem photovoltaic cell are higher top-cell current densities (CGSe₂ has demonstrated up to 20mA/cm² saturated photocurrent) and less shadowing of an underlying cell. The PEC cell eliminates the need for a TCO window and grids an overlying PV cell would, transmitting more light to an underlying PV cell (as shown in Figures 1B and 1C). It should be kept in mind that for a PEC device in operation, the sun will still have pass through a container wall (quartz, glass, plastic) as well as the electrolyte itself. Piece-wise simulation using characterization data from J-V measurements, UV/visible spectrophotometry, quantum efficiency (QE) measurements of PV cells and incident photon-to-current efficiency (IPCE) measurements of PEC cells were carried out using load line analysis to determine a theoretical operating point before going forward with device integration.

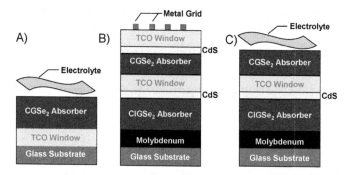

Figure 1: Schematics (not to scale) of (A) Glass/TCO/CGSe$_2$ bare PEC device (no voltage bias), (B) Glass/Mo/CIGSe$_2$/CdS/TCO/CGSe$_2$/CdS/TCO/Grid tandem PV device and (C) a Glass/Mo/CIGSe$_2$/CdS/TCO/CGSe$_2$ hybrid-tandem PEC/PV device. A comparison of (B) and (C) show how a tandem-hybrid device has fewer layers, minimizing shading and fabrication costs.

EXPERIMENTAL DETAILS

Fabrication

Three different alloys of the I-III-VI$_2$ material class were utilized for this study. All thin-film absorbers were fabricated in our custom 5-source vacuum evaporation system (5 sources being Cu, In, Ga, Se and a fifth source for either Na or Ag) utilizing 2 crystal monitors and an Inficon Guardian Co-Deposition Controller using electron impact emission spectroscopy (EIES) to monitor individual effusion rates during deposition. In all materials, a 3-stage deposition method was employed using end point detection to obtain a slightly Cu-poor film (I/III~0.7-0.9 ratio where I=Cu, III=In,Ga) [8]. CGSe$_2$ for PEC performs better when I/III is closer to 0.7 (more Cu-poor) for PEC applications while films for PV applications (usually CIGSe$_2$) perform better with I/III closer to but not higher than 0.9 [6]. CGSe2 films for PEC cells were deposited on glass/SnO$_2$:F (FTO) substrates (TEC 15 from Hartford Glass) with an average thickness of 700nm at 450°C (highest temperature). CIGSe$_2$ (1.8μm) and CISe$_2$ (2.2 μm) films for PV cells were deposited on glass/Mo substrates. PV absorbers were finished with a CdS buffer (chemical bath deposition), a ZnO/ITO window (sputtered) and a Ni/Ag grid (evaporated) for a final glass/Mo/CIGSe$_2$/CdS/ZnO/ITO/Ni/Ag structure.

Monolithic stack devices utilized PV devices *without* grids, with CGSe$_2$ deposited directly on the window layer with the procedure described above. The cells were then scribed with a razor blade, wired, and epoxied to electrically isolate the underlying PV cell from the electrolyte.

Characterization

Illumination was provided by an Oriel 1000W solar simulator calibrated by photodiodes tuned to the bandgaps of corresponding materials. All PEC cells were characterized using 0.5M H_2SO_4 electrolyte in a 2- or 3- terminal arrangement using a RuO_2 thin-film counter electrode unless otherwise stated. QE and IPCE measurements were done using a PV Measurements Inc. custom system and UV/vis measurements were carried out using a PerkinElmer Lambda 2 Spectrometer. A Hitachi S-4800 FESEM was used to take micrographs and an included Oxford INCA X-Act EDS system was used to obtain elemental composition data.

DISCUSSION

Hybrid Device Simulation

While the $CISe_2$ ($E_g{\sim}1.0eV$) PV cells should have produced a higher current and lower voltage by absorbing lower energy photons, the cells produced around the same current (29-32mA/cm^2) as the $CIGSe_2$ ($E_g{\sim}1.1eV$ by design using Ga/(In+Ga)~0.3 ratio [1]). Moreover, $CIGSe_2$ had an open circuit voltage V_{OC}=542mV (vs. only a V_{OC}=386mV for $CISe_2$) leading to a much higher efficiency of η=11.69% in $CIGSe_2$ (vs. only η=6.40% for $CISe_2$). A QE analysis of the $CISe_2$ showed large losses in the lower energy spectrum which is likely a fabrication issue.

Continuing the investigation of candidate tandem components with $CGSe_2$ paired with $CIGSe_2$, QE measurements were taken of the $CIGSe_2$ PV cell both with and without optical filtering through a full $CGSe_2$ electrode (glass/FTO/CGSe$_2$). The results of these measurements are shown in Figure 2 where the filtered QE area is hatched and labeled "CIGSe$_2$". The IPCE of the $CGSe_2$ device is also shown (area filled and labeled "CGSe$_2$" in Figure 2) and as expected, its absorption drops off as the $CIGSe_2$ absorption begins to rise (both around the $CGSe_2$ bandgap of 1.65eV~750nm). In the background is the standard AM1.5$_G$ spectrum for reference.

Integrating the quantity of the QE values times the standard AM1.5$_G$ solar spectrum (converted from watts to coulombs of photons) results in the theoretical current that particular cell will produce when illuminated by sunlight [9,10]. Integrating the curves bordering the solid areas shown in Figure 2, the $CGSe_2$ PEC cell shows an expected current of 12.21mA/cm^2 (data taken at -2V vs. a platinum counter electrode, roughly agreeing with JV data for this sample) while the $CIGSe_2$ PV cell shows an expected short circuit current (QE taken at V=0) of 8.26mA/cm^2 (in good agreement with JV data showing J_{sc}=7.96mA/cm^2 under calibrated, simulated AM1.5$_G$ illumination shaded by a $CGSe_2$ cell). In a tandem cell integrated with these two devices, Kirchhoff's circuit laws maintain that the current will be limited by the lowest current in the system (here the $CIGSe_2$ PV cell) and the voltages for corresponding current values will be added together.

Because JV measurements are taken as a function of voltage, a series-equivalent curve cannot be immediately calculated. However, a very good approximation can be obtained by adding voltages where the PEC and PV currents are matched. This calculation is shown in Figure 3 with the theoretical hybrid PEC/PV photocathode (solid line) as well as the as-measured $CGSe_2$ curve (dotted line). As expected, there is a voltage shift towards positive (meaning less required voltage bias to achieve water-splitting) and a saturation in current due to the current mismatch between the two cells, limiting current to that of the lowest value (the CIGSe$_2$ PV cell, Jsc=7.96mA/cm^2).

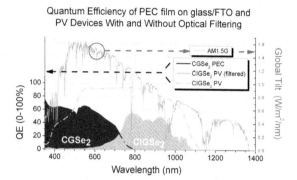

Figure 2: QE curves of CIGSe$_2$ both as-produced (dotted line) and optically filtered (hatched area labeled "CIGSe$_2$") by a CGSe$_2$ photocathode (glass/FTO/CGSe$_2$). The IPCE curve of the CGSe$_2$ photocathode with an operating voltage bias applied is shown as the solid area labeled "CGSe$_2$".

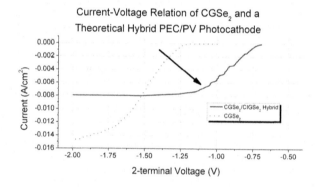

Figure 3: Current-Voltage relationship of the as-measured CGSe$_2$ photocathode (dotted line) and the theoretical PEC performance of a hybrid PEC/PV (solid line) photocathode utilizing the CIGSe$_2$ PV cell indicated in Figure 2. Comparison shows an as-expected tradeoff between a lower required voltage bias (desired) and a decreased device photocurrent.

In regards to I-III-VI$_2$ materials used for photocathodes in water-splitting, required voltage reduction is the highest priority so this simulation does show that this system (CGSe$_2$/CIGSe$_2$) is at least feasible [7]. Even with the reduction in current due to current mismatch, this theoretical device could potentially produce an appreciable amount of hydrogen. Also of note, this simulation is based on the QE curve in Figure 2 which includes the absorption of the substrate of the PEC electrode (glass/FTO) as well as the metallic grid on the PV cell shading it. Monolithic stacking of these devices will eliminate those absorption losses and increase the limiting current. Of course, an optimal system would pair PEC and PV devices with *matched* currents, requiring a PEC top cell with a higher bandgap. In the I-III-VI$_2$ material class, this could be accomplished by the addition of sulfur (to fully or partially replace selenium) [7].

Hybrid Device Fabrication

Verification through simulation that a voltage reduction can be achieved while maintaining appreciable photocurrent warrants advancement towards realizing a hybrid PV/PEC device. For monolithic integration, three matters of concern have been identified thus far. Upon PEC testing of monolithic hybrid photocathodes (there is no 3rd terminal to access the PV cell independently), every cell (after showing varying degrees of photoactivity) would corrode in seconds before peeling off the underlying substrate. Electron microscopy revealed that there were areas of the hybrid device where the top CGSe$_2$ layer seemed to fuse through the window TCO layer to the CIGSe$_2$ below it, as can be seen in Figure 4. This fusion is suspected of shunting the cell leading to corrosion once either light or external voltage is applied.

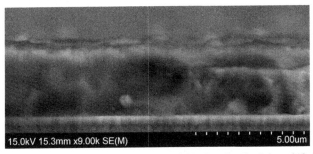

15.0kV 15.3mm x9.00k SE(M)   5.00um

Figure 4: SEM micrograph showing areas where the top CGSe$_2$ layer has fused through the TCO to the CIGSe$_2$ below it (at left side of figure). This shunt is suspected to begin a corrosion process once light or external bias is applied leading to peeling of the entire top layer. The majority of the films do, however, show intact layers as shown in the right portion of the figure.

While a small imperfection will cause a corrosion process to propagate across the entire surface of the film, the majority of the cross sectional area showed intact layers. Energy dispersive spectroscopy (EDS) did, however, reveal a decrease in In concentrations after either heating or hybrid fabrication, shown in Figure 5. Heat tests were carried out with typical deposition parameters of vacuum and temperature profiles. The resultant I/III ratio shows that the films were too Cu rich, which often leads to shunting or other degradations in PV performance.

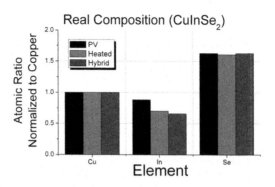

Figure 5: Elemental composition data (normalized to Cu) obtained through EDS analysis showing a decrease in indium content in the bulk absorber when it is heated either in vacuum or in the deposition of a top PEC layer for a hybrid device. Selenium content appears to be stable.

Last, but not least, SEM analysis showed degradation in the crystallinity of the $CGSe_2$ top cell. While standalone PEC films take on a scaly morphology, the $CGSe_2$ on top of the PV cell window showed a more amorphous or cauliflower structure which could be due to either the difference in ITO (on PV cell) and FTO (typical substrate) or the morphology of the window itself being grown on a rough surface instead of smooth glass.

CONCLUSION

In this study, simulated models from real data suggest that $CGSe_2$ PEC cells can be combined with either $CIGSe_2$ or possibly $CISe_2$ to produce a hybrid PEC/PV photocathode to split water. While this tandem cell will not provide *all* the voltage required to split water, it can theoretically reduce the potential enough to make photoelectrochemical water-splitting more economically feasible. This simulation also points out the large current mismatch between $CGSe_2$ and $CIGSe_2$, which could be rectified by the incorporation of sulfur (in place of selenium) to raise the bandgap of the top cell, transmitting more light to an underlying PV cell.

A monolithically stacked device will also remove many losses these simulations incorporated like a glass/TCO layer as well as grid shadowing of the PV cell. Fabrication issues regarding integration of the constituent parts into one, monolithic device include mechanical failure of the TCO window when heated for the 2nd absorber deposition and altered crystallinity of the top PEC cell. A thicker TCO layer may improve both issues.. It should be noted that this hybrid cell represents a reduction in failure modes compared to tandem PV/PV devices because it eliminates the buffer/window/grid layers a PV top cell would require.

Indium loss (and consequent Cu-enrichment) may be mitigated by depositing excess In to begin with, expecting some to re-evaporate upon the 2nd absorber deposition. The deposition temperature (450°C here) can also be lowered further; so far, lower deposition temperatures have been improving $CGSe_2$ PEC performance.

ACKNOWLEDGEMENTS
This research is supported by the U.S. Department of Energy Award Number DE-FG36-07GO17105 A000. We would also like to thank Tina Carvalho at the University of Hawaii Biological Electron Microscope Facility for her patience and assistance completing the SEM and EDS analyses presented here.

REFERENCES

1. M. Bär et al., J. Appl. Phys. **96,** 3857 (2004).
2. M. Schmid et al., PV Direct **1,** 10601 (2010).
3. J. Shewchun et al., J. Appl. Phys. **50,** 6978 (1979).
4. M.E. Beck et al., Thin Solid Films **361-362,** 130 (2000).
5. M.R. Balboul et al., J. Vac. Sci. Technol. A **20**(4), 1247 (2002).
6. B. Marsen et al., Mater. Res. Soc. Symp. Proc. **974,** 0974-CC09-05 (2007).
7. J. Kaneshiro et al., Sol. Energ. Mat. Sol. Cells. **94,** 12 (2010).
8. I. Repins, et al., Prog. Photovolt: Res. Appl. **16,** 235 (2008).
9. T. Markvart and L. Castañer, *Practical Handbook of Photovoltaics: Fundamentals and Applications*, Elsevier Science Inc., New York, NY, 10010), pp. 77-78.
10. W. Shockley and H.J. Queisser, J. Appl. Phys. **32,** 510 (1961).

Mater. Res. Soc. Symp. Proc. Vol. 1324 © 2011 Materials Research Society
DOI: 10.1557/opl.2011.842

Identification of Impurity Phases in Cu_2ZnSnS_4 Thin-film Solar Cell Absorber Material by Soft X-ray Absorption Spectroscopy

M. Bär,[1] B.-A. Schubert,[1] R.G. Wilks,[1] B. Marsen,[1] Y. Zhang,[2] M. Blum,[2,3] S. Krause,[2] W. Yang,[3] T. Unold,[1] L. Weinhardt,[4] C. Heske,[2] and H.-W. Schock[1]

[1]Solar Energy Research, Helmholtz-Zentrum Berlin für Materialien und Energie GmbH (HZB), Hahn-Meitner-Platz 1, D-14109 Berlin, Germany
[2]Department of Chemistry, University of Nevada, Las Vegas, 4505 Maryland Pkwy., Las Vegas, NV 89154, U.S.A.
[3]Advanced Light Source (ALS), Lawrence Berkeley National Laboratory, 1 Cyclotron Rd., Berkeley, CA 94720, U.S.A.
[4]Exp. Physik VII, Universität Würzburg, Am Hubland, D-97074 Würzburg, Germany

ABSTRACT

The composition of Cu_2ZnSnS_4 thin-film solar cell absorbers was varied to induce the formation of secondary impurity phases. For their identification, the samples have been investigated by Cu L_3 and S $L_{2,3}$ soft x-ray absorption (XAS) spectroscopy. We find that Cu L_3 XAS is especially sensitive to the presence of copper sulfides as well as copper oxides and/or changes in the electron configuration, suggesting a basis for future studies of the surface, defect, and interface characterization of similar samples. Additionally, it is shown that the S $L_{2,3}$ absorption data can be used as a very sensitive probe of the variations in the prevalence of S-Zn bonds in the near-surface region of the investigated samples.

INTRODUCTION

The two most important factors in the commercial development of photovoltaic (PV) technologies are the achieved solar cell device efficiency (on laboratory and large scale) and the costs per power in an industrial mass production. On these grounds, various thin-film PV technologies already successfully compete with the current state-of-the-art Si-wafer based solar cells. Solar cells based on the Cu-chalcopyrite alloy system $(Cu(In_{1-X}Ga_X)(S_YSe_{1-Y})_2)$ reach efficiencies in excess of 20 % [1] in the laboratory, while for CdTe-based thin-film devices (record efficiency: 16.5 % [2]) industrial large-scale manufacturing costs below 0.80 US\$/W_p [3] are reported. Thus thin-film PV devices are able to convert sunlight into electrical energy at the same performance level as polycrystalline Si-wafer based devices [4] at potentially lower production costs. Due to the limited availability of some $Cu(In_{1-X}Ga_X)(S_YSe_{1-Y})_2$ absorber constituents, concern exists that high material expenses will jeopardize the promise of lower industrial-scale production costs for this potentially higher performing thin-film PV technology. As a result, kesterites used as thin-film solar cell absorbers have attracted much attention in the recent past [5], and efficiencies of up to 9.6 % [6] have already been reported for $Cu_2ZnSn(S,Se)_4$-based photovoltaic devices.

In order to reach the next level of solar cell performance, a detailed insight into the chemical and electronic structure of kesterites is necessary. However, so far the electronic structure of the $Cu_2ZnSn(S,Se)_4$ material system has only been thoroughly investigated in theoretical studies [7], while detailed experimental analyses are still lacking. One reason for this deficit is, presumably, the complex (structural) phase diagram of the Cu-Zn-Sn-S-Se system [8] and related problems with the deposition of high quality material free of impurity phases and/or mixed crystal systems. Rapid advancements in device efficiency and commercialization can only occur once reproducible processes for the deposition of kesterite thin-film solar cell absorbers free of secondary impurity phases are developed. For this study, a series of samples was produced in which the bulk composition of Cu_2ZnSnS_4 ("CZTS") kesterites was deliberately varied in a range relevant for solar cell devices in order to induce the formation of secondary impurity phases. The chemical structure of these samples was then investigated by energy-dispersive x-ray spectroscopy (EDS) and soft x-ray absorption spectroscopy (XAS). EDS probes the elemental composition of the entire volume of the investigated thin films, while XAS provides complementary element-resolved and chemically-sensitive information based on the selective excitation of specific core electrons into the unoccupied energy levels of the material. Comparing the chemical structure derived from non-destructive EDS with the more surface-sensitive XAS measurements can thus give "depth-resolved" information about the investigated CTZS thin films.

EXPERIMENT

Sample preparation

Polycrystalline CZTS thin films of different bulk compositions were deposited on molybdenum-coated soda lime glass using thermal coevaporation of copper, tin, and zinc sulfide source materials. The substrate was heated up to 550°C before the copper, tin, sulfur, and zinc sulfide sources were opened simultaneously. A more detailed description of the deposition process can be found in Ref. [9]. This process results in closed and homogeneous CTZS thin films of approximately 1.5 μm thickness [9]. Scanning electron microscopy and x-ray diffraction measurements indicate the formation of secondary $Cu_{2-x}S$ ($0 \leq x \leq 1$) phases on top of the Cu-rich processed CTZS thin films [9]. To obtain more comparable results all samples were thus etched in aqueous KCN solution for 3 min at room temperature prior characterization. After KCN etching (done at HZB), the samples intended for XAS analysis were immediately transferred to a nitrogen purged glovebox, where they were sealed in desiccated plastic bags for transport to the ALS.

Cu_2S, CuS, ZnS, and SnS_2 powder references (purity > 99%) pressed in indium foil were also characterized for comparison. Since the copper sulfides are prone to oxidation in ambient air, the powder containers were opened and the chemicals mounted in a nitrogen-purged glovebag connected directly to the load lock chamber of the XAS analysis system. However note that while the CuS powder was packed under argon in a glass ampoule ("ultra dry", Alfa Aesar), the Cu_2S reference was provided in a conventional glass container (presumably packed under air). The CZTS/Mo thin-film samples were unpacked and mounted in air (together with the ZnS and SnS_2 powders) shortly prior to the transfer into the analyses system.

Sample characterization

The bulk chemical composition was determined by EDS using a LEO440 scanning electron microscope with tungsten hairpin cathode using an energy-dispersive Si(Li) detector. The characteristic L emission lines of zinc and tin and the K emission lines of copper and sulfur were used for the quantitative composition analysis. The acceleration voltage was tuned between 12–20 kV adapting the information depth to the thickness of the investigated CZTS thin films in order to avoid an interference of the S K with the Mo L emission line from the substrate.

The XAS measurements were conducted at Beamline 8.0.1 of the Advanced Light Source using the soft x-ray fluorescence (SXF) endstation [10]. The base pressures of the load lock system and the SXF analyses chamber were better than 1×10^{-6} and 1×10^{-9} mbar, respectively. The Cu L_3 XAS spectra were measured in total fluorescence yield mode (TFY) using a channeltron positioned in front of the sample and in electron yield mode (TEY) using the sample current, the S $L_{2,3}$ XAS spectra were measured in partial fluorescence yield (PFY) mode using the permanently installed SXF spectrometer as detector. For this purpose, the SXF spectrometer was optimized for the S $L_{2,3}$ emission and gated such that the Rayleigh line did not influence the measurement. All XAS spectra were normalized to the incoming monochromatic photon flux, which was monitored via the photocurrent of an Au-covered mesh positioned between the last optical element of the beamline and the sample. The energy scales of the Cu and S $L_{2,3}$ XAS spectra were calibrated according to Refs. [11] using Cu and CdS references, respectively. As mentioned above, while EDS probes the composition of the entire volume of the CZTS thin films, the 1/e information depth for Cu L_3 TFY and TEY (S $L_{2,3}$ PFY) XAS is approx. 200 nm and 10 nm (20 nm) [12], respectively, and hence can be considered near-surface bulk sensitive.

RESULTS AND DISCUSSION

Table 1 summarizes the bulk composition of the investigated samples as revealed by EDS. The most extreme deviation from the nominal Cu:Zn:Sn:S = 2:1:1:4 (i.e., 25:12.5:12.5:50 at%) composition can be found for sample (a). However, samples (b) – (f) also show significant variations in the Zn/Sn as well as Cu/(Zn+Sn) ratios. Excluding sample (a), the Zn/Sn ratio changes from 0.74 to 1.01 between samples (b) and (f) with a Cu/(Zn+Sn) ratio varying around the nominal stoichiometry: $0.90 \leq 1 \leq 1.08$. Solar cell devices based on the CZTS absorber material represented by samples (a) – (f) result in efficiencies between 0.2 and 4.1 % [9].

Table 1: Composition of the investigated CZTS/Mo samples as derived by EDS and power conversion efficiencies of corresponding thin-film solar cell devices.

Sample	Composition [at%]				Zn/Sn	Cu/(Zn+Sn)	η [%]
	Cu	Zn	Sn	S			
(a)	12.0	33.8	5.8	48.4	5.83	0.30	0.8
(b)	27.0	10.6	14.3	48.1	0.74	1.08	0.2
(c)	24.8	12.6	14.1	48.5	0.89	0.93	3.3
(d)	26.0	12.4	13.5	48.1	0.92	1.01	3.9
(e)	24.3	13.5	13.6	48.6	0.99	0.90	2.7
(f)	26.3	12.9	12.8	48.0	1.01	1.02	4.1

Because secondary $Cu_{2-x}S$ phases are expected to form during the deposition process [9], we first focus on whether they are exclusively formed at the CZTS surface and furthermore whether such phases are reliably removed by KCN etching. Fig. 1(A) compares the Cu L_3 XAS spectra of the CZTS/Mo samples to the measured absorption of the Cu_2S and CuS powder references as well as literature TEY XAS data for Cu_2S, CuS, Cu_2O, and CuO (digitized from Ref. [13]). Note that while the absorption spectra of the CZTS thin films were normalized to the edge jump between 927.5 and 939 eV, the XAS spectra of the Cu_2S and CuS powder references as well as the literature data were normalized to the maximum of the prominent absorption features at 931 and 932 eV, respectively. For samples (a) – (d) and the powder references, both the TFY (solid, lines, filled dots) as well as the TEY (dashed lines, open dots) XAS spectra are presented. While the Cu L_3 absorption of the CuS agrees well with corresponding literature data [13], the

93

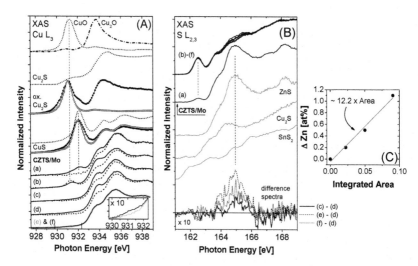

Figure 1 (A): Cu L$_3$ XAS spectra of the CZTS/Mo samples (a) – (f) [lines] compared to the absorption spectra of copper sulfide powder references [dots] and TEY literature data (digitized from Ref. [13]) for Cu$_2$S, CuS, Cu$_2$O, and CuO [dotted/dashed lines]. For CTZS thin films (a) – (d) and the powder references the TFY [solid lines, filled dots] and the TEY [dashed line, open dots] XAS spectra are shown. The inset shows an enlarged view of the TFY absorption spectra (energy range indicated by short vertical lines) of thin film (e) [grey] and of the highest-performing CZTS material (f) [black]. **(B):** S L$_{2,3}$ XAS spectra of the CZTS/Mo samples (a) – (f) compared to the absorption spectra of binary compound powder references: Cu$_2$S, ZnS, and SnS$_2$. At the bottom, the tenfold enlarged difference spectra (c)-(d), (e)-(d), and (f)-(d) are presented. **(C):** The change of the absolute amount of Zn in the samples [relative to the content in sample (d)] is presented versus the difference spectra area (between 164 and 166 eV) shown in (B). The fit of the linear relation is also shown.

measured XAS data of the Cu$_2$S powder reference do not agree with the Cu$_2$S literature spectrum – the emergence of the peak at 931 eV suggests the presence of CuO. Though much less pronounced, note that also the literature spectra of Cu$_2$O and Cu$_2$S exhibit intensity in that energy range, which was also attributed to the presence of CuO impurities [13]. Above 933 eV, the shape of the measured Cu$_2$S TFY spectrum resembles the literature Cu$_2$S spectrum. The more surface sensitive TEY measurement, however, is in good agreement with the literature CuO absorption data. Thus, we conclude that a (partially) oxidized Cu$_2$S powder reference was analyzed, and as a result the sample is henceforth denoted oxidized Cu$_2$S ("ox. Cu$_2$S"). Note that the shoulder in the respective TFY spectrum at 932 eV might indicate CuS as additional impurity. The Cu L$_3$ XAS spectra of the CZTS/Mo samples (d) – (f) look very similar to each other, and the difference between the TEY and TFY data can mainly be attributed to the well-known self-absorption effect in the TFY XAS spectra. In contrast, for the thin film (a), it is observed that while the TFY spectrum shows a small peak at the same energetic position as the main absorption feature of CuS, the TEY spectrum resembles spectra (d) – (f). This suggests the presence of CuS in the bulk of sample (a). Most likely, this phase was also present at the surface of CZTS structure (a), but here it could successfully be removed by the KCN etch step. For sample (b), we find an absorption feature with an energy similar to the main Cu L_3 XAS feature of the oxidized Cu_2S powder and CuO (approx. 931 eV). Whether this can be interpreted as oxidized Cu or is indicative for unfilled Cu 3d states ($\rightarrow$ $3d^9$ configuration) in CZTS sample (b) cannot be unambiguously determined. Note that the predominant electron ground state configuration for Cu in formally monovalent (bivalent) compounds as Cu_2S, Cu_2O, or CZTS (CuS or CuO) is expected to be $3d^{10}$ ($3d^9$) [13,14]. However, for CuS there is only spectroscopic evidence for $3d^{10}$ states [14,15]. The first explanation (i.e. a surface oxidation) is, however, in agreement with the higher intensity of that absorption feature observed for the more surface sensitive TEY XAS measurement. Close inspection of the direct comparison of the Cu L_3 XAS spectra of thin-film sample (e) and (f) (inset of Fig. 1(A)) reveals a slight intensity variation in the same energy range. Interpreting the higher spectral intensity around 931 eV for the highest-performing CZTS material (f) as an indication for a slight CuO formation would be surprising and thus we speculate that this intensity modulation might be ascribed to a slightly different (beneficial?) electron configuration: (increased) presence of unfilled Cu 3d states despite the nominal $3d^{10}$ character of CZTS (see discussion above). In Fig. 1(B), the S $L_{2,3}$ absorption spectra of the CZTS/Mo samples (a) – (f) are compared with the reference spectra of the binary compounds Cu_2S, ZnS, and SnS_2,. All spectra were normalized to the edge jump between 161 and 169 eV. As expected from the EDS data of sample (a) presented in Table 1, the S $L_{2,3}$ XAS of structure (a) does not resemble the absorption of samples (b) – (f), which are very similar. The good agreement with the S $L_{2,3}$ absorption of the ZnS reference however indicates that the S in sample (a) predominantly forms bonds to Zn (also confirmed by XRD). The small peak in the spectrum of (a) at 162.5 eV coincides with the most prominent contribution to the S $L_{2,3}$ absorption of samples (b) – (f) and can be attributed to hybridized *antibonding* Sn 5s – S 3p states which dominate the bottom of the CZTS conduction band [16]. Hence a minor amount of S in a CZTS environment can also not be excluded for sample (a). A close inspection of the S $L_{2,3}$ spectra (b) – (f) reveals that the absorption spectra deviate in the range between 164 and 166 eV. The prominent absorption features in that energy range of the ZnS and Cu_2S reference spectra suggests that a varying contribution of S-Zn and/or S-Cu bonds could explain the observed intensity modulation. In order to analyze this in more detail, the following difference spectra were computed: (c)-(d), (e)-(d), and (f)-(d), and are shown (tenfold enlarged) in the lower part of Fig. 1(B). Since sample (b) was already identified to significantly deviate in the chemical structure (see above), the S $L_{2,3}$ XAS spectra (b) was not included in the difference analysis. The most prominent feature of all difference spectra is a broad peak centered at 165 eV that increases in the following order: (c)-(d) $\rightarrow$ (f)-(d) $\rightarrow$ (e)-(d). Because it energetically coincides with the most intense absorption feature of the ZnS reference spectrum, we attribute this to an increasing contribution of S-Zn bonds. In order to verify this interpretation, the areas below the difference spectra were quantified via integration between 164 and 166 eV. Taking the absolute Zn content of sample (d) (12.4 at%, see EDS data in Table 1) as a point of reference, one can compute the absolute difference in the amount of Zn (Δ Zn) in the CZTS samples (c), (e), (f) to 0.2, 1.1, and 0.5, respectively. The resulting plot of Δ Zn over the quantified area of the difference spectra in Fig. 1(C) shows a linear correlation, confirming the origin of the observed intensity variation as indicative of differences in the prevalence of S-Zn bonds. The linear fit Δ Zn $\approx$ 12.2 $\times$ Area can thus be used to determine the change in the absolute Zn concentration directly from the difference spectrum.

SUMMARY & CONCLUSION

The chemical structures of CZTS thin films with different bulk stoichiometries were investigated using Cu L_3 and S $L_{2,3}$ XAS. Leaving aside sample (a), which – based on the EDS analysis – exhibited a pronounced deviation from the nominal Cu : Zn : Sn : S = 2 : 1 : 1 : 4 composition, the Cu L_3 absorption data revealed a different Cu chemical environment for sample (b) than in samples (e)-(f). Furthermore, the S $L_{2,3}$ XAS data provided a very sensitive probe for the amount of S-Zn bonds in the investigated samples. In this case, the observed spectral intensity variations were directly linked to the observed variation in the absolute amount of Zn in the CZTS thin films, opening a route for the quantitative evaluation of the collected S $L_{2,3}$ spectra. For the CZTS samples where no significant spectral deviations could be observed [→ samples (c) – (f)], the corresponding solar cells have decent power conversion efficiencies between 2.7 and 4.1%. Future experiments will show in which cases the intensity difference in the Cu L_3 spectra of CZTS thin films in the energetic range of the most prominent absorption feature of CuO is a mere fingerprint for Cu oxidation or an indication for a different (beneficial?) electron configuration.

ACKNOWLEDGMENTS

M. Bär and R.G. Wilks acknowledge the financial support by the Helmholtz-Association (VH-NG-423). The ALS is funded by the Department of Energy, Basic Energy Sciences, Contract No. DE-AC02-05CH11231.

REFERENCES

1. I. Repins et al., Prog. Photovolt. Res. Appl. **16**, 235 (2008); and P. Jackson et al., Prog. Photovolt. Res. Appl., published online, DOI: 10.1002/pip.1078235 (2011).
2. X. Wu et al., Proc. 17th Eu-PVSEC, 22–26 October, Munich, Germany, 2001, 995.
3. Investor Relations, First Solar, Inc., http://investor.firstsolar.com, 04/10/2011.
4. M.A. Green et al., Prog. Photovolt. Res. Appl. **19**, 84 (2011).
5. T.M. Friedlmeier et al., Ternary and Multinary Compounds. Proc. 11th Int. Conf., ICTMC-11, Salford, UK, Vol. **152**, 345 (1998); T. Tanaka et al., J. Phys. Chem. Sol. **66**, 1978 (2005); H. Katagiri et al., Appl. Phys. Express **1**, 041201 (2008).
6. T.K. Todorov et al., Adv. Mater. **22**, 1 (2010).
7. S. Chen et al., Appl. Phys. Lett. **94**, 041903 (2009); J. Paier et al., Phys. Rev. B **79**, 115126 (2009); M. Ichimura et al., Jpn. J. Appl. Phys. **48**, 090202 (2009); C. Persson, J. Appl. Phys. **107**, 053710 (2010).
8. G. Moh, Chemie der Erde **34**, 1 (1975); I.D. Olekseyuk et al., J. Alloys Compd. **368**, 135 (2004).
9. B.-A. Schubert et al., Prog. Photovolt. Res. Appl. **19**, 93 (2011).
10. J.J. Jia et al., Rev. Sci. Instrum. **66**, 1394 (1995).
11. L. Weinhardt et al., Phys. Rev. B **79**, 165305 (2009); M. Bär et al., Appl. Phys. Lett. **93**, 244103 (2008).
12. Attenuation lengths from B. L. Henke et al., At. Data Nucl. Data Tables **54**, 181 (1993); http://wwwcxro.lbl.gov/optical_constants/atten2.html.
13. M. Grioni et al., Phys. Rev. B **39**, 1541 (1989).
14. S.W. Goh et al., Minerals Engineering **19**, 204 (2006) and Refs. therein.
15. G. van der Laan et al., J. Phys. Chem. Solids **53**, 1185 (1992).
16. M. Bär et al., unpublished.

Mater. Res. Soc. Symp. Proc. Vol. 1324 © 2011 Materials Research Society
DOI: 10.1557/opl.2011.844

Kesterites and Chalcopyrites: A Comparison of Close Cousins

Ingrid Repins[1], Nirav Vora[1], Carolyn Beall[1], Su-Huai Wei[1], Yanfa Yan[1], Manuel Romero[1], Glenn Teeter[1], Hui Du[1], Bobby To[1], Matt Young[1], Rommel Noufi[1]
[1]National Renewable Energy Laboratory, 1617 Cole Blvd., Golden, CO 80401, U.S.A.

ABSTRACT

Chalcopyrite solar cells based on $CuInSe_2$ and associated alloys have demonstrated high efficiencies, with current annual shipments in the hundreds of megawatts (MW) range and increasing. Largely due to concern over possible indium (In) scarcity, a related set of materials, the kesterites, which comprise Cu_2ZnSnS_4 and associated alloys, has received increasing attention. Similarities and differences between kesterites and chalcopyrites are discussed as drawn from theory, depositions, and materials characterization. In particular, we discuss predictions from density functional theory, results from vacuum co-evaporation, and characterization via x-ray diffraction, scanning electron microscopy, electron beam-induced current, quantum efficiency, secondary ion mass spectroscopy, and luminescence.

INTRODUCTION

Chalcopyrite solar cells have progressed from first proof of concept on melt-grown crystals [1] to high-efficiency thin-film laboratory devices [2,3] to manufacturing presently in the hundreds of MW range. Manufacturing volumes are expected to push into the gigawatt (GW) range in coming years due to the combination of relatively high efficiency (compared to CdTe or α-Si) and potentially low processing costs (compared to single-crystal Si).

As production volumes increase, concern has arisen as to whether limits in the world supply of In will restrict the amount of chalcopyrite photovoltaics that can be manufactured at low cost. While the current cost of In adds only 1 to 10 ¢/Watt to the price of module manufacturing [4], a tenfold increase in In price would be problematic for the necessary sub-$1/W manufacturing cost. Estimates of when In scarcity will become important to module prices range in $CuIn_{1-x}Ga_xSe_2$ (CIGS) manufacturing volumes from 4 GW/yr to over 100 GW/yr [4,5,6,7]. If estimates of 4 GW/yr are correct, In scarcity could impact the industry within 10 years. If estimates of over 100 GW/yr are correct, and multiple technologies continue to be close in cost, In scarcity may not impact the photovoltaic industry.

To mitigate possible future effects of In scarcity, it has been proposed that CIGS can be replaced by $Cu_2ZnSn(S,Se)_4$ in the kesterite structure, in which every two group III (In or Ga) atoms in chalcopyrite structure are replaced by a Zn (group II) and Sn atom (group IV), thus maintaining the octet rule. To date, solar cells with up to 9.7% efficiency [8] have been made using these In-free kesterite absorbers. In our work with the co-evaporation of chalcopyrites and kesterites, we have observed both similarities and differences between absorbers made using these closely related crystal structures. Several aspects of this comparison are discussed in the sections below.

PREDICTIONS FROM DENSITY FUNCTIONAL THEORY

Structural, electronic, and defect properties of both kesterites and chalcopyrites have been investigated using density functional theory [9] (DFT). These examinations have revealed both similarities and differences between kesterites and chalcopyrites.

Positions of conduction and valence band edges have been calculated for both kesterites and chalcopyrites [10,11,12], as shown in Table I for $CuInSe_2$ (CISe), $Cu_2ZnSnSe_4$ (CZTSe), $CuInS_2$ (CIS), and Cu_2ZnSnS_4 (CZTS). In the table, positions of conduction and valence band edges are shown relative to vacuum, where the relative band offsets have been calculated from DFT, and the experimental value of the electron affinity for CISe, 4.6 eV [13], is taken from experiment. For the selenides and sulfides, the kesterite band gaps are very similar to (slightly less than) the corresponding chalcopyrite band gaps. Furthermore, each material maintains a reasonable alignment with the CdS conduction band at -4.3 eV, explaining the success of the CdS/i-ZnO/ZnO:Al buffer and window layers in finishing both chalcopyrite and kesterite devices. It should be noted that for both the chalcopyrite and the kesterite, other crystal structures with the same stoichiometry exist. For I-III-VI$_2$ compounds, a chalcopyrite or CuAu-like structure may form. For I$_2$-II-IV-VI$_4$ compounds, a kesterite, stannite, or primitive mixed CuAu structure may form. However, DFT calculation predicts that the chalcopyrite and kesterite structures are the lowest energy configuration for the materials considered here [14].

Table I. Calculated band gaps, valence band edges, and conduction band edges (relative to vacuum) for selenide and sulfide chalcopyrites and kesterites.

Material	E_g (eV)	E_v (eV)	E_c (eV)
CISe (chalcopyrite)	1.04	-5.64	-4.60
CZTSe (kesterite)	1.00	-5.56	-4.56
CIS (chalcopyrite)	1.53	-5.92	-4.39
CZTS (kesterite)	1.50	-5.71	-4.21

The stability of single-phase stoichiometric material has been calculated as a function of the chemical potential of constituent atoms for both chalcopyrites [15] and kesterites [16]. For $CuInSe_2$, single-phase material can occur for Cu chemical potentials $-0.4 < \mu_{Cu} < 0$ eV and the corresponding In chemical potential $-0.4 < \mu_{Cu} < -1.8$ eV. For CZTS, single-phase material occurs only at a much smaller chemical potential space around ($\mu_{Cu} = -0.20$ eV, $\mu_{Zn} = -1.23$ eV, and $\mu_{Sn} = -0.50$ eV). It is thus predicted that constraints on forming stoichiometric CZTS are more strict for kesterites than for chalcopyrites.

Chemical defects impose further limits on compositions or reaction paths that may yield high-efficiency solar cells. Because there is one more element in the kesterite structure than in the chalcopyrite structure, there are a larger number of possible intrinsic defects for CZTS and related compounds than for CIS. Formation energies and transition energy levels within the gap have been calculated for the 13 dominant defects that do not involve the unlikely situation of a cation on an anion lattice site or vice versa [11]. Of these 13 defects, five have relatively low formation energy for Cu-poor kesterite. These five low-formation-energy defects are, in order of increasing formation energy: copper on zinc site (denoted Cu_{Zn}), copper vacancy (denoted V_{Cu}), Zn_{Sn}, V_{Zn}, and Cu_{Sn}. The formation energy of Cu_{Zn} is significantly lower than that of V_{Cu} and Zn_{Sn}, predicting that Cu_{Zn} should be the dominant intrinsic defect for CZTS. The predicted abundance of Cu_{Zn} is a different situation than for the chalcopyrite, where V_{Cu} is the lowest formation energy defect [15] and is responsible for the p-type doping. The Cu_{Zn} acceptor level is predicted to lie about 0.1 eV above the valence band maximum (VBM), as opposed to the shallower acceptor V_{Cu} (0.02 or 0.03 eV above the VBM in the kesterite or chalcopyrite, respectively). Since kT ≈ 0.025 eV at room temperature, the fraction of the dominant acceptors ionized may be significantly different in the kesterite than in the chalcopyrite. For both the chalcopyrite [17] and the kesterite [11], n-type doping is predicted to be unlikely due to the high formation energy of the donor defects and low formation energy of acceptor defects.

For both chalcopyrites and kesterites, low-formation-energy defects with energy levels deep within the gap exist, but may be largely benign due to the formation of defect complexes. For CISe, the most important defect for carrier recombination is predicted to be In_{Cu} due to both its low formation energy and its predicted level 0.34 eV below the CBM [15]. However, the electronic activity of this defect is to some extent passivated by the formation of $[2V_{Cu} + In_{Cu}]$ defect complexes [15]. Of the five low-formation-energy defects previously mentioned for CZTS, Cu_{Sn} is expected to be the most active recombination center, with an energy level close to mid-gap, ~0.6 eV above the VBM [11]. As is the case for CISe, DFT calculations predict that charge-compensated defect complexes such as $[Cu_{Sn} + Sn_{Cu}]$ are likely to form and passivate deep levels in the kesterites [11]. The dominant defect complex in CZTS is $[Cu_{Zn} + Zn_{Cu}]$. However, unlike in CISe, where the formation of $[2V_{Cu} + In_{Cu}]$ defect complexes leads to charge-separated α and β phases [17], the formation of $[Cu_{Zn} + Zn_{Cu}]$ does not have this possibly beneficial effect observed in chalcopyrite compounds.

The effects of alloying CZTS and CZTSe have also been investigated using DFT calculations [18]. It is shown that the enthalpy of formation for the $CZT(S_{1-x}Se_x)$ is very small, indicating that the mixed-anion alloys are highly miscible, and that the cations maintain the same ordering preferences as in pure kesterites. The band gaps of the random alloy decrease with Se content almost linearly. The conduction band down-shift contributes more to the gap decrease than the valence band up-shift. The shift of the band edges makes the $CZT(S_{1-x}Se_x)$ alloys with high Se concentration easier to be doped both n-type and p-type. The balance between the band gap size and the doping ability may therefore be important in determining the optimal alloy composition to achieve high-efficiency $CZT(S_{1-x}Se_x)$-based solar cells.

FILM DEPOSITION

Co-evaporation of CIGS films at the National Renewable Energy Laboratory (NREL) and their incorporation into device structure has been previously discussed in detail [19]. To make kesterite films and devices, identical procedures were followed, except that In and Ga shot were replaced in the evaporator by Zn and Sn shot, and appropriate optical filters (202.55 nm for Zn and 284.0 nm for Sn) were substituted for the Ga and In filters in the electron impact emission spectrometer rate monitor. For devices discussed in this study, anti-reflective coating was *not* applied.

The two types of CZTSe recipes used for films in this study are shown in Figure 1. In all depositions, rates and temperatures were stabilized, then a substrate shutter was opened at time = 0. Substrate temperature was held constant in the range of 470-500°C throughout the deposition. Temperature variations within this range are noted in the text when discussing different samples. Deposition temperatures are relatively low, due to decomposition of the kesterite at temperatures typically used during CIGS formation [20]. Depositions lasted about 20 minutes. In all cases, Se and Sn were supplied in overpressure throughout the deposition, similar to the conditions for Se flux during CIGS deposition. Cu and Zn fluxes were supplied in two variations. In the first variation, shown in Figure 1a, Cu and Zn fluxes are supplied throughout the deposition period. Thus, the Cu/Zn and Cu/(Zn+Sn) atomic ratios in the film are constant with time. This type of deposition is analogous to "one-stage" depositions used for CIGS. In the second type of recipe, shown in Figure 1b, the Cu rate is increased so that Cu/Zn and Cu/(Zn+Sn) can be greater than one during most of the deposition, and yet the final film can be Cu-poor. We observe during such depositions that the Cu-rich or Cu-poor character of the film is indicated by emissivity signature [21], in much the same way as for CIGS films, presumably due to the effects of excess copper selenide.

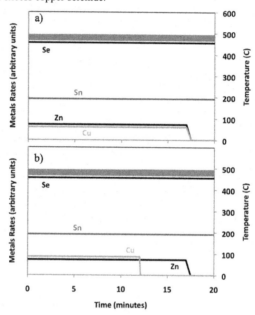

Figure 1. Schematic representation of recipes used for samples in this study. For CIGS, these types of recipes are commonly called a) "one-stage" and b) "two-stage."

A summary of the composition measurements for the samples described in this paper is shown in Table II. Compositions are shown in terms of atomic ratio. Film compositions were measured by x-ray fluorescence (XRF) calibrated against inductively coupled plasma optical emission spectrometry (ICP-OES). The table also shows the recipe type for each sample (either one-stage or two-stage) and the figure in which data for each sample is presented. X-ray diffraction, Raman spectroscopy, energy dispersive spectroscopy, and resistance measurements were used to confirm that the films in this paper are kesterites, with the occasional ZnSe precipitate in Zn-rich samples. The results and application of these phase identification techniques to co-evaporated CZTSe will be discussed in a separate publication [22].

Table II. Summary of compositions and recipe types for films used in this study.

Sample	Zn/Sn	Cu/(Zn+Sn)	Cu/Sn	Recipe type	Figure in which sample is characterized
M3214	1.2	1.0	2.15	One-stage	Figure 3a
M3250	1.35	0.8	1.9	One-stage	Figure 3b
M3253	1.0	0.85	1.7	Two-stage	Figure 2a
M3251	1.2	0.95	2.0	Two-stage	Figure 2b
M3272	1.1	0.95	2.0	Two-stage	Figure 2c
M3268	1.2	0.85	1.8	Two-stage	Figure 2d
M3257	1.05	0.9	1.9	Two-stage	Figure 4
M3215	0.75	1.2	2.1	One-stage	Figure 5
M3244	1.1	0.9	1.9	Two-stage	Figure 6, Figure 8

MORPHOLOGY

Parallels are observed in the morphology versus vacuum growth conditions of chalcopyrites and kesterites. These similarities include increasing grain size with both temperature and occurrence of a Cu-rich growth period.

Grain size in CIGS films increases with temperature [23]. CZTSe films in this study have shown a similar trend. Figure 2 shows scanning electron micrograph (SEM) images of CZTSe films grown under nearly identical conditions, except for increasing substrate temperature. The compositions for these films, as measured by x-ray fluorescence (XRF), are also listed in Table I. Feature size in the SEM images increases with substrate temperature. Note that the temperature range in Figure 2 is considerably narrower than that which has been examined for CIGS [23]. The films of Figure 2b and Figure 2d, with the highest Zn/Sn ratios in the sample set, begin to show segregation of ZnSe, which is evident as the bright spots near the Mo in the SEM images. Energy dispersive spectroscopy (EDS), used to examine these bright spots while the films are mounted in the electron microscope, confirms that the bright spots are mostly Zn and Se. For the films in this study, the increase in grain size with temperature did not yield an increase in device performance.

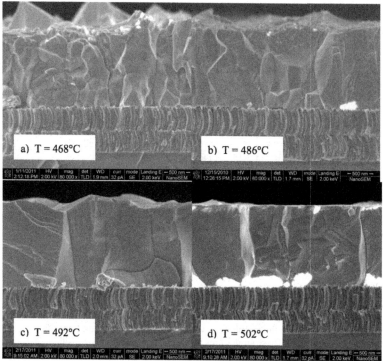

Figure 2. SEM cross-sections of CZTSe films grown under nearly identical conditions, except for varying substrate temperatures of a) 468°C, b) 486°C, c) 492°C, or d) 502°C.

Chalcopyrite films have been documented to exhibit increased grain sizes when a period of Cu-rich (i.e., Cu/(In+Ga) >1) growth is included in the deposition [24,25,26]. This improved grain growth has been associated with a flux recrystallization of chalcopyrite grains by a liquid Cu_xSe phase. In the present work, kesterite films have been observed to exhibit a similar dependence on Cu-rich vs. Cu-poor growth. Figure 3 shows two kesterite films made under nearly identical conditions. For both, the substrate temperature was about 490°C, and a one-stage recipe was utilized. However, for the film in Figure 3a, Cu rate was chosen such that the Cu/(Zn+Sn) ratio just exceeds one. For the film in Figure 3b, Cu rate was chosen such that the Cu/(Zn+Sn) ratio was always below one. The film grown under Cu-rich conditions exhibits considerably larger grain size. Other researchers have also observed large grain sizes under Cu-rich growth conditions [27,28]. For the film in Figure 3a, a sodium cyanide etch was used to remove excess copper selenide prior to acquiring the SEM image.

The relationship between Cu ratio and grain size for CZTSe films is likely multi-variate and should be investigated more fully. It has been shown for CIGS that the amount of benefit obtained via a Cu-rich growth period is a function of deposition temperature [23]. Furthermore, a marked dependence of CZTSe grain size on temperature is demonstrated in Figure 3. Variations in composition other than Cu excess may also have an effect on morphology.

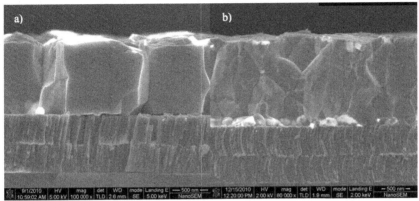

Figure 3. SEM cross-section of CZTSe films grown at the same temperature and constant deposition fluxes, except Cu rates were adjusted to produce only a) Cu-rich growth or b) Cu-poor growth.

COMPOSITION PROFILES

Figure 4 shows Auger emission spectroscopy (AES) profiling of a CZTSe film that yielded 4.4% devices. In Figure 4a, the entire film thickness (1.5 µm) is shown. Composition is fairly constant throughout the bulk. Figure 4b expands the near-surface region of the same data, plotted this time in terms of atomic ratio, in order to examine trends in the constituent elements without seeing effects from several atomic percent oxygen or carbon at the surface. It is evident in Figure 4b that, while the amount of Zn and Sn in the film is constant, the amount of Cu drops considerably near the surface. The two minutes of sputtering time over which this drop occurs corresponds to roughly 70 nm. Cu-poor surfaces are also observed in CIGS [29].

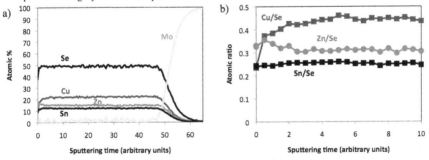

Figure 4. AES profile of a) entire film and b) near-surface region.

Secondary ion mass spectrometry (SIMS) profiling was used to compare the amount of Na in standard CIGS with that in the CZTSe films. Na is known to be important in CIGS devices for increasing free carrier density and decreasing recombination [30,31]. Positive secondary data are shown compared for CIGS (thin dashed lines) and CZTSe (thick solid lines) in Figure 5. For both films, Se makes up ~50% of the lattice, and Se counts are normalized to 10^3 to allow for comparison of amounts of other elements. The CZTSe film is only half the thickness of the CIGS film, so the SIMS data is horizontally offset so that the back contact (i.e., the area of Mo signal increase) lines up for both films. The 1 μm CZTSe thickness was chosen to minimize requirements on elemental diffusion and final film conductivity during these initial investigations. The striking feature of Figure 5 is that CZTSe film contains much less Na than the CIGS, as might be expected from the necessarily lower deposition temperature, since diffusion through the Mo is the main barrier to Na inclusion in CIGS [32]. Determining from device results whether the optimum amount of Na in CZTSe is the same as that in CIGS should be an area of future examination.

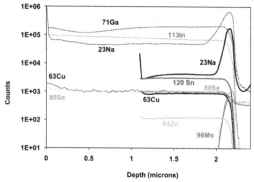

Figure 5. SIMS data comparing CIGS (thin dashed lines) and CZTSe (thick solid lines).

ELECTRO-OPTICAL CHARACTERISTICS

Both CIGS and CZTSe samples have been mapped using electron beam-induced current (EBIC) and cathodoluminescence (CL). The methodology and apparatus used for luminescence microscopy has been described in detail in earlier publications [33]. EBIC measurements are performed in a field-emission scanning electron microscope with electron beam energies of 2-3 keV. The secondary electron image is acquired prior to the EBIC image at extremely low electron beam currents (20-50 pA), while EBIC images are acquired at beam currents of 100-200 pA. No degradation of the EBIC signal is observed under these low excitation conditions. For CL measurements, electron beam energy is 15 keV and saample temperature is 20 K.

To examine the effect of morphology on carrier collection, cross-sectional SEM images and cross-sectional electron beam induced current (EBIC) maps were compared between CIGS and CZTSe. Figures 6a and b show typical SEM and EBIC images for a ~16%-efficient CIGS device. The dotted lines in Figures 6a and b are benchmarks to indicate the same physical location on the sample in each image. Also shown in Figure 6b is a linescan, i.e., an x-y representation of the image brightness at the location where the linescan is drawn. The salient feature of Figures 6a and b is that the response is basically constant as the electron beam moves away from the junction, although grain boundaries and an occasional grain show reduced response. Figures 6b and c show the same type of characterization for a 4.1% CZTSe device. Carriers are collected only within tenths of a micron of the junction, and this response length is nonuniform over the sample. The data in Figure 6 imply that improving collection length is extremely important in early work in improving CZTSe devices.

This conclusion is consistent with quantum efficiency (QE) data taken on progressively improving devices in this study. The QE data are shown in Figure 7 and show increasing red collection with improving efficiency. A 20.0%-efficient CIGS device is also shown in Figure 7 for comparison. The minimal red loss (i.e. flat QE from ~700-900 nm) is apparent in the CIGS device, in strong contrast with the kesterite devices shown. Also evident is the difference in band gap between the CIGS and CZTSe (wavelength at which QE approaches 0) and the lack of anti-reflective coating on the CZTSe devices (~5% reduction in maximum QE).

CL images of the surface of CZTSe films were also compared with those from CIGS. Several aspects of this comparison are consistent with the high recombination center density indicated by the EBIC and the QE. First, the intensity of the luminescence is about two orders of magnitude lower for the CZTSe than for the CIGS. Second, the materials show a different behavior in emitted photon energy with increasing CL current. This comparison is shown in Figure 8. CL as a function of excitation current for CIGS of 1.15 eV band gap is shown in figure 8a, and that for CZTSe is shown in Figure 8b. Figure 8b was taken on the same material used to make the 4.1% device of Figure 6. For both CZTS and CIGSe, the spectrum consists of a broadband luminescence associated with transitions between donor and acceptor defect bands. The electronic states participating in the formation of these bands are deeper for CZTSe (~0.2 eV below the band gap) than CIGS (~0.1 eV below the band gap). The spectrum shows a significant blueshift with the external excitation (measured by the electron beam current) for both chalcopyrites and kesterites. This blueshift is due to the filling of the bands while increasing the excitation. We observe that (i) the blueshift is much more pronounced for CIGS than CZTSe, and (ii) there is a saturation of the bands at high excitation in CIGS (confirmed by the change in the shape of the peak with the e-beam current for CIGS), which is not observed in CZTSe. Thus, not only are the dominant intrinsic defects leading to the formation of these bands deeper in energy for CZTSe when compared to CIGS but, based on these measurements, the density of these defects is much higher in CZTSe.

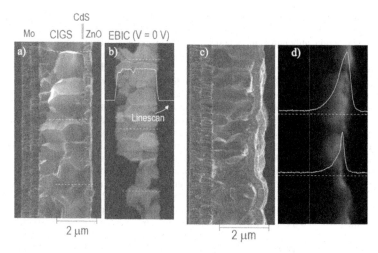

Figure 6. a) SEM and b) EBIC cross-sections for CIGS; and c) SEM and d) EBIC cross-sections for CZTSe.

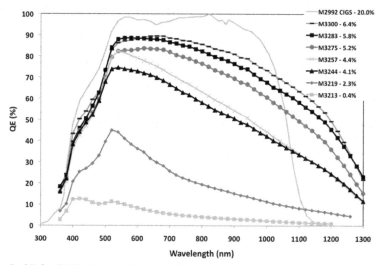

Figure 7. QE for CZTSe devices of various efficiencies.

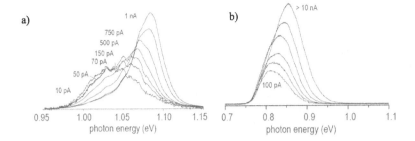

Figure 8. CL spectrum as a function of CL current for a) CIGS and b) CZTSe.

CONCLUSIONS

A comparison of chalcopyrites and kesterites yields both significant similarities and differences. Theory predicts similar band edge energies, p-type doping, formation of benign defect complexes, and miscibility of alloys. However, theory also predicts differences: a smaller single-phase composition space and a different dominant intrinsic defect. During vacuum co-evaporation, the chalcopyrites and kesterites can be made using nearly identical equipment, recipes, and endpoint detection. However, co-evaporation of the kesterites requires Sn (not just Se) overpressure and lower temperatures. Kesterites and chalcopyrites both show an increase in grain size with temperature and with a Cu-rich growth period. A Cu-poor surface is observed in both types of film, although the universality of this observation has not been tested for the kesterites. The kesterite deposition recipes used in this study result in much less Na in the absorber than in optimized CIGS films, as might be expected from the necessarily lower deposition temperature. The kesterite films in this study show evidence of short diffusion lengths, in sharp contrast with good CIGS devices. This evidence of increased activity of recombination centers is evident in cross-sectional EBIC, QE, and CL.

ACKNOWLEDGMENTS

This work was supported by the U.S. Department of Energy under Contract No. DE-AC36-08-GO28308 with the National Renewable Energy Laboratory. The authors thank Jeffrey Blackburn and Jian V. Li for helpful discussions.

REFERENCES

1. S. Wagner, J. L. Shay, and P. Migliorato, *Appl. Phys. Lett.* **25**(8), 434-435 (1974).
2. P. Jackson, D. Hariskos, E. Lotter, S. Paetel, R. Wuerz, R. Menner, W. Wischmann, M. Powalla, *Prog. Photovoltaics*, DOI: 10.1002 (2011).

3. I. Repins, M.A. Contreras, B. Egaas, C. DeHart, J. Scharf, C.L. Perkins, B. To, and R. Noufi, *Prog. Photovoltaics* **16**, 235–239 (2008).
4. V. Fthenakis, *Renew. Sust. Energ. Rev.* **13**, 2746–2750 (2009).
5. T. Sullivan, W.H. Kuo, P. Karayan, M. LoCascio, M. Holman, S. Udupa, B. Hilman, A. Stuk, "Solar State of the Market Q3 2008: The Rocky Road to $100 Billion," *LRSI-SMR-08-02*, (Lux Research, 2008).
6. A. Feltrin, A. Freundlich, *Renew. Energ.* **33**, 180–185 (2008).
7. B.A. Andersson, *Prog. Photovoltaics* **8**, 61-76 (2000).
8. T. K. Todorov, K. B. Reuter, D. B. Mitzi, *Adv. Mater.* **22**, 1-4 (2010).
9. G. Kresse, J. Furthmuller, *Comp. Mater. Sci.* **6**(1), 15-50 (1996).
10. S. Chen, X.G. Gong, A. Walsh, S.H. Wei, *Appl. Phys. Lett.* **94**, 041903 (2009).
11. S. Chen, J.H. Yang, X.G. Gong, A. Walsh, S.H. Wei, *Phys. Rev. B* **81**, 245204 (2010).
12. S.H. Wei, A. Zunger, J. Appl. Phys. (6), 3846-3856 (1995).
13. N. Romeo, *Jpn. J. Appl. Phys.* **19-3**, 5-13, (1980).
14. D.S. Su, S.H. Wei, *Appl. Phys. Lett.* **74**(17), 2483-2485 (1999).
15. S. B. Zhang, S.H. Wei, A. Zunger, H. K. Yoshida, *Phys. Rev. B* **57**(16), 9642-9656 (1998).
16. S. Chen, X.G. Gong, A. Walsh, S.H. Wei, *Appl. Phys. Lett.* **96**, 021902 (2010).
17. S.H. Wei, B. Zhang , *J. Phys. Chem. Solids* **66**, 1994 (2005).
18. S. Chen, A. Walsh, J.-H. Yang, X.-G. Gong, L. Sun, P.-X. Yang, J.-H. Chu, and S.-H. Wei, *Phys. Rev. B* **83**, 125201 (2011).
19. A.M. Gabor, "The conversion of (In,Ga)$_2$Se$_3$ thin films to Cu(In,Ga)Se$_2$ for application to photovoltaic solar cells," *PhD thesis*, (University of Colorado, 1995), pp. 42-49.
20. A. Weber, R. Mainz, H.W. Schock, *J. Appl. Phys.* **107**, 013516 (2010).
21. J. Kessler, J. Scholdstrom, L. Stolt, *IEEE Phot. Spec. Conf.* **28**, 509-512(2000).
22. N. Vora et al, in progress.
23. W.N. Shafarman, J. Zhu, *Mater. Res. Soc. Symp. P.* **668**, H2.3.2-H2.3.6 (2001).
24. A.M. Gabor, J.R. Tuttle, D.S. Albin, M.A. Contreras, R. Noufi, *Appl. Phys. Lett.* **65** (2), 198-200 (1994).
25. R.A. Mickelsen, W.S. Chen, Y.R. Hsiao, V.E. Lowe, *IEEE Transactions on Electron Devices* **ED-31** (5), 542-546 (1984).
26. J.R. Tuttle, M.A. Contreras, M.H. Bode, D. Niles, D.S. Albin, R. Matson, A.M. Gabor, A. Tennant, A. Duda, R. Noufi, *J. Appl. Phys.* **77** (1), 153-161 (1995).
27. B.A. Schubert, B. Marsen, S. Cinque, T. Unold, R. Klenk, S. Schorr, H.W. Schock, *Prog. Photovoltaics* **19** (1), 93-96 (2011).
28. T. Tanaka, A. Yoshida, D. Saiki, K. Saito, Q. Guo, M. Nishio, T. Yamaguchi, *Thin Solid Films* **518**, S29-S33 (2010).
29. S.H. Han, F.S. Hasoon, A.M. Hermann, D.H. Levi, *Appl. Phys. Lett.* **91**, 021904 (2007).
30. J.E. Granata, J.R. Sites, *World Conf. Photo. Sol. Energ. Conv.* **2**, 604-607 (1998).
31. D. Rudmann, "Effects of sodium on growth and properties of Cu(In,Ga)Se$_2$ thin films and solar cells," *PhD thesis*, (Swiss Federal Institute of Technology, 2004).
32. M.B. Zellner, R.W. Birkmire, E. Eser, W.N. Shafarman, J.G. Chen, *Prog. Photovoltaics* **11**, 543-548 (2003).
33. M.J. Romero, M.A. Contreras, I. Repins, C.S. Jiang, M.M Al-Jassim, *Mater. Res. Soc. Symp. P.* **1165**, 419-424 (2010).

Mater. Res. Soc. Symp. Proc. Vol. 1324 © 2011 Materials Research Society
DOI: 10.1557/opl.2011.1058

Impact of thickness variation of the ZnO:Al window layer on optoelectronic properties of CIGSSe solar cells

Jan Keller[1], Martin Knipper[1], Jürgen Parisi[1], Ingo Riedel[1], Thomas Dalibor[2], Alejandro Avellan[2]
[1]Energy and Semiconductor Research Laboratory, D-26111 Oldenburg, Germany
[2]AVANCIS GmbH & Co. KG, Otto-Hahn-Ring 6, D-81739 Munich, Germany

ABSTRACT

We studied the thickness variation of equally doped ZnO:Al films used as conductive window layer in Cu(In,Ga)(Se,S)$_2$ (CIGSSe) thin film solar cells. The IV-characteristics of solar cells with window layer thickness of d_1=200nm exhibit a strong enhancement of the short-circuit current density J_{SC} (ΔJ_{SC} = 3mA/cm^2) as compared to samples with module-like ZnO:Al-film thickness (d_2=1200nm). Accordingly, the quantum efficiency reveals the spectral regimes where the J_{SC}-gain occurs. Moreover, current-voltage measurements reveal that the cells with thicker ZnO:Al exhibit slightly decreased open circuit voltage V_{OC}. This finding can be assigned to a decreased net-doping density N_A, which appears to be introduced by additional heat flux during the longer process time required for deposition of thicker ZnO:Al films. However, the improved efficiency of solar cells with thinner window layer comes along with an increase of the series resistance (R_S) by almost a factor of 2, which will have consequences for the series connection of elements in a module. XRD-diffractograms and SEM cross-section imaging suggest that the enhanced R_S in cells with thin ZnO:Al is not exclusively related to the thickness but is also due to a reduced (002)-texture and an elongated lateral charge carrier pathway.

INTRODUCTION

Predicting low production-costs and competitive module efficiencies, thin film technologies enter the photovoltaic market and gain increasing attention. Besides CdTe and thin film silicon, chalcopyrite-based concepts, i.e. Cu(In$_{1-x}$Ga$_x$)(S$_y$,Se$_{1-y}$)$_2$, are most promising to compete with wafer technologies. The highest lab-scale efficiency reported for co-evaporated CIGSe exceeds 20% [1]. Being likely for thin film technologies, the performance transfer to large areas appears to be difficult as reflected by the reduced module efficiencies (η=13-14% champion modules [1-2]). One approach to enhance the module performance is to modify the front-contact, usually made of transparent conducting oxides (TCO). The main requirements for the TCO are a good lateral conductivity and a high optical transmission in the range where the CIGSSe absorbs. The first issue is realized via heavy doping and employment of thick films to achieve sufficiently low sheet resistance. In contrast, the TCO should be thin in order to achieve a high transmission in short wavelength range whereas low doping density is expected to reduce parasitic absorption by free carrier absorption (FCA). Furthermore, reflection losses can be avoided by surface-texturing, for instance by using nanorod-shaped arrays [3] or applying a surface-etching [4]. Most commonly, aluminum-doped ZnO (ZnO:Al) is used as a low-cost TCO in inorganic solar cells and photovoltaic modules. In this work we discuss the impact of film thickness variation on the optical and electrical characteristics of state-of-the-art CIGSSe solar cells.

EXPERIMENT

For optical and structural characterization ZnO:Al was applied on glass substrates by DC magnetron sputtering with three different thicknesses at the same doping level. A 1200nm thick layer (d_2) commonly used in present modules serves as a reference sample. The samples with 200nm (d_1) and a 1600nm (d_3) ZnO:Al should reveal advantages in decreasing and increasing the TCO-thickness. To study the impact of the TCO thickness on the cell and module performance, CIGSSe cells (nominally same absorber) with identical ZnO:Al thickness-variation have been produced via a two-step process including RTP. Here, glass was used as a substrate, molybdenum as back contact, CdS as the buffer layer and a thin i-ZnO layer for passivation. After processing the substrate was sectioned down to single cells (A_{cell}=1.65cm²) by mechanical scribing. Optical transmission and reflection spectra of the ZnO:Al-specimen on glass have been recorded by a Cary 500 spectrometer. Surface and cross-section investigations were performed using an Agilent 5420 AFM and additionally a FEI SEM/FIB-system. The crystallographic studies were performed with an XRD spectrometer (X'Pert PANalytical) using the Cu-Kα line. Electrical characterization of the completed cells was done by current-voltage measurements under standard test conditions (STC), spectral response and capacitance-voltage measurements. For this a class A solar simulator, a home-built EQE-setup and a HP 4194A impedance analyzer were used, respectively.

DISCUSSION

Electro-optical characterization

Figure 1(a) shows the STC-approximated IV-characteristics of representative CIGSSe cells including ZnO:Al window layers with the thickness variation under discussion. As expected the short-circuit current density J_{SC} increases by ΔJ_{SC}=3.5mA/cm² when TCO-thickness is reduced from 1600nm to 200nm. The cell main parameters are summarized in table 1.

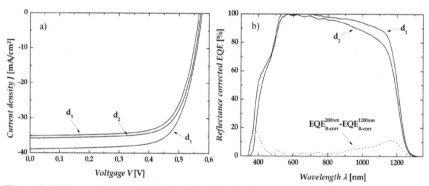

Figure 1. IV-characteristics of CIGSSe cells with varied TCO thickness (d_1=200nm, d_2=1200nm, d_3=1600nm) (a) and reflectance-corrected quantum efficiency for d_1 and d_2 (b). The dashed line shows the difference spectrum.

Apart from the gain in J_{SC} we observe some little improvement in open-circuit voltage (ΔV_{OC}=8mV). The fill factor (FF) remains quite stable for all samples leading to a total enhancement of the cell efficiency by $\Delta\eta$=1.8%-points for a TCO-thickness reduction from d_3=1600nm to d_1=200nm, mainly due to improvements in short-circuit current density. To reveal the spectral regimes of the J_{SC}-enhancement the reflectance-corrected EQE (EQE$_{R-corr}$) of samples d_1 and d_2 are compared in figure 1(b) (sample d_3 not shown for clarity). The difference in EQE$_{R-corr}$ is expected to reflect the different parasitic absorption losses in the respective ZnO:Al layers. Figure 1(b) shows that EQE$_{R-corr}$ of sample d_1 exceeds the values of sample d_2 at wavelengths below 500nm and above 700nm, which is explained by the reduced thickness. While in the first regime band-to-band absorption is the dominant loss mechanism, free charge carrier absorption (FCA) dominates in the low-energy part. The difference in EQE$_{R-corr}$ continuously increases with increasing wavelengths (dashed line) while approaching the plasmonic wavelength λ_p (>1500nm), where the FCA reaches its maximum. Calculating the resulting current densities yields a gain of ΔJ_{SC}=0.5mA/cm^2 due to less interband absorption (λ<500nm) and ΔJ_{SC}=1.3mA/cm^2 caused by reduced FCA losses in the thinner TCO (λ>700nm). For the case of low band-gap CIGSSe compositions the loss in short-circuit current is dominated by FCA. An approach to minimize these losses while keeping the conductivity compatible could be to exchange the dopant (e.g. boron - ZnO:B), where the carrier mobility is increased apparently [5].

Table 1. Main solar cell parameters derived from IV-measurements shown in Figure 1(a)

TCO thickness [nm]	200	1200	1600
V_{OC} [mV]	574	571	566
J_{SC} [mA/cm^2]	38.9	35.8	35.4
FF [%]	71.7	72.1	71.5
η [%]	16.1	14.7	14.3

The net acceptor density N_A profiles of the absorbers of the different samples were calculated from CV-profiles and are presented in figure 2(a). We observe a distinct trend to a lower net doping density for cells with larger TCO thickness. Since the absorbers have been processed identically and are therefore nominally the same the effect is anticipated to be introduced by the TCO sputtering process (differences due to chemical bath deposition of CdS are excluded). To investigate the impact of an extended heat treatment during the sputtering process for thicker TCOs we annealed sample d_1 for 20 minutes at 200°C under argon atmosphere. This annealed sample shows a (even larger) similar decrease of the apparent doping density by a factor of 3. This drop has a direct impact on V_{OC} as shown in figure 2(b) where the IV-curves of sample d_1 for the as grown state and after temperature annealing are compared. The loss in open-circuit voltage is ΔV_{OC}=10mV and agrees well with the 8mV difference observed for d_1 and d_3. We conclude that even at quite moderate temperatures being present during TCO deposition (around 200°C) the electronic properties of the absorber may be substantially altered and could lead to reduced cell performance.

The decreased cross section area of the sample with the thin TCO (d_1) leads to a decreased sheet conductivity. This directly increases the series resistance R_S of the solar cell which might influence the fill factor. From dark IV-analysis of 16 cells (not shown) we calculated the mean R_S-values of the discussed cells applying a method described elsewhere [6].

R_S increases from sample d_2 to d_1 by a factor of ~2 ($R_{S,1200nm}=0.51\Omega cm^2$ and $R_{S,200nm}=0.90\Omega cm^2$) while thicker TCO does not yield further reduction ($R_{S,1600nm}=0.55\Omega cm^2$).The increased R_S was

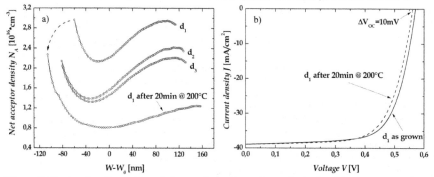

Figure 2. Doping density profiles of discussed samples and sample d_1 after 20 minutes annealing at 200°C (a). IV-characteristics of sample d_1 in the as-grown state and after annealing at 200°C (b).

found to have only a minor effect on the performance of a single cell (FF remains constant). However, in a series-connected module where all ohmic losses will add up this might lead to a drop of FF and could therefore reduce the efficiency. In the next part we discuss whether the increase in R_S is exclusively a thickness effect of the TCO or if there are additional structural origins.

Crystal- and microstructure investigations

For determination of the crystal structure and growth direction of doped zinc oxide films X-ray diffractograms have been recorded under different tilting angles. All ZnO:Al-films on glass exhibit the typical hexagonal lattice structure (wurtzite) as shown in figure 3(a). Below the XRD-spectra the expected peak-intensity ratios are shown for intrinsic ZnO powder [7]. It is obvious that the specimen exhibit a heavily textured growth, with pronounced (002) and (004) reflections. The preferred growth direction for ZnO:Al along the c-axis has been reported earlier [8]. Beside the (002) peak the (103) peak becomes more and more distinct for thinner TCO films, which indicates reduced (002)-texture. Comparative XRD-measurements on fully processed CIGSSe-cells confirm this trend, where ZnO:Al shows up in the same lattice-structure (not shown). In figure 3(b) the respective intensity ratios of the (103) and (002) peaks are illustrated. It can be clearly observed, that for both "substrates" the texture becomes less unique for thinner samples. The inset in figure 3(b) shows the pole figures of the (002) peak detected at a fixed angle of $2\theta=34.4°$ with varied tilting angles of the sample. All samples show a distinct peak near to the centre which directly confirms a (002)-texture while the peak (and therefore the degree of (002) texturing) becomes sharper for thicker TCOs. In the case of sample d_1 an additional maximum appears at approximately 30° tilting angle that corresponds to an admixture of the (103) texture. It seems that the growth of ZnO:Al is initially a mixture of a (002) and (103) orientation while the (002) texture becomes more and more dominant for thicker layers (sharpest (002) peak for d_3). Lee et. al reported that the film resistivity of aluminum-doped ZnO is inversely proportional to the degree of (002) grain orientation [9]. This gives rise to the assumption of decreased sheet conductivity in thinner ZnO:Al films. We conclude that the increased R_S observed in samples with thin TCO can be assigned also to this structural effect.

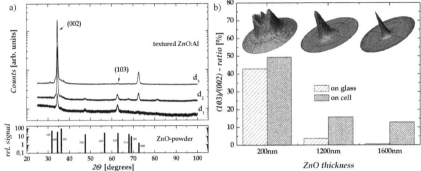

Figure 3. XRD spectra of ZnO:Al films on glass and intensity ratio of ZnO powder (a). Peak-intensity ratios (103)/(002) on glass and on the processed cells are shown in (b). The inset displays the pole figures of the (002) reflections for ZnO:Al/glass.

The surface of the processed films is rather rough and wavy. AFM measurements proved a maximum surface waviness, i.e. the peak-to-valley distance, of about 1100nm (not shown). Figure 4 shows the SEM cross section images of the investigated samples prepared by focused ion beam structuring. It is obvious that the ZnO:Al inherits the wavy topology of the CIGSSe-absorber. For TCO thickness below the maximum waviness, in this case sample d_1 (a), the charge carriers cannot laterally propagate along a straight path (like it is possible for both thicker TCOs in (b) and (c)) to the next cell. Evaluating several FIB cross sections (>10) the elongation in charge carrier pathway in samples with thin TCO window layer was estimated to about 3-4%, which will have some negative effect on sheet conductivity. We conclude that the increase in series resistance observed in sample d_1 can be ascribed to three mechanisms: (1) reduced TCO cross section (2) less (002) textured growth and (3) topology effect causing elongated charge carrier pathway.

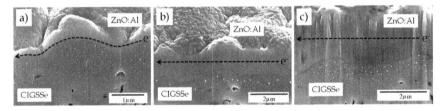

Figure 4. Cross sections of CIGSSe solar cells with the three TCO thicknesses prepared by focused ion beam (thickness increases from left to right). Dashed arrows demonstrate possible lateral charge carrier pathways.

CONCLUSIONS

In this study the impact of thickness variation from conventional "module size" (1200nm) to very thin (200nm) and thick (1600nm) ZnO:Al-layers has been investigated. CIGSSe solar cells with reduced TCO thickness exhibit a substantial enhancement of the short-circuit current density of about 3mA/cm^2. It was confirmed by reflectance-corrected EQE measurements that this gain is mainly due to reduced free charge carrier absorption in the TCO for wavelengths >700nm. The gain due to less interband absorption for photon energies exceeding the band gap of ZnO:Al (3.32eV) was less than the half. We conclude that in chalcopyrite solar cells with low band gap absorbers (e.g. CIGSSe) FCA should be addressed by fine-tuning the TCO thickness. In addition to the increased J_{SC} also V_{OC} is slightly increased in cells with thinner TCO (ΔV_{OC}=8mV). We have shown that this phenomenon is introduced by unintentional absorber annealing during the extended process time required for deposition of thick TCO films. The additional introduction of heat leads to a decreased apparent doping density in the space charge region of the absorber, followed by a slight drop of V_{OC}. Thinning the TCO to 200nm leads to a twofold increase of R_S in the solar cells (not greater because R_S is also determined by other parameters that give a constant resistance offset). We have shown that this is probably not exclusively a matter of a decreased TCO cross section but also due to structural changes. XRD-measurements confirmed that in the initial stages of the ZnO:Al growth the (002)-texture which is beneficial for conductivity is less distinct and becomes more and more pronounced with increasing layer thickness. Furthermore the wavy topology of the RTP-processed cells induces an elongated charge carrier pathway for TCO thickness below approximately 1100nm (about 3-4% for d_1=200nm). Accordingly, the increased series resistance is not only related to the electrical film properties but further depends on the lattice structure and absorber topology.

ACKNOWLEDGMENTS

Funding of the "EWE Nachwuchsgruppe Dünnschicht-Photovoltaik" by the EWE AG Oldenburg, Germany is gratefully acknowledged.

REFERENCES

1. M. A. Green, K. Emery, Y. Hishikawa, and W. Warta, *Progress in Photovoltaics: Research and Applications* (2010).
2. N. G. Dhere, *Solar Energy Materials and Solar Cells* **95**, pp. 277-280 (2010).
3. D. Kieven, J. Chen, R. Klenk, T. Rissom, Y. Tang, and M. C. Lux-steiner, *Progress in Photovoltaics: Research and Applications* **18**, pp. 209-213 (2010).
4. O. Kluth, B. Rech, L. Houben, S. Wiedera, G. Schöpe, C. Benekinga, H. Wagner, A. Löffl, H.W. Schock, *Thin Solid Films* **351**, pp. 247-253 (1999).
5. Y. Hagiwara, T. Nakada, A. Kunioka, *Sol. Energy Mater. Sol. Cells* **67**, pp. 267–271 (2001).
6. S. S. Hegedus and W. N. Shafarman, *Progress in Photovoltaics: Research and Applications* **12**, pp. 155-176 (2004).
7. R. Heller and J. McGannon, *J. Appl. Phys.* **21**, p. 1283 (1950)
8. N. P. Dasgupta, S. Neubert, W. Lee, O. Trejo, J.-R. Lee, and F. B. Prinz, *Chemistry of Materials*, pp. 4769-4775 (2010).
9. J.-H. Lee and B.-O. Park, *Thin Solid Films* **24**, pp. 94-99 (2003).

Mater. Res. Soc. Symp. Proc. Vol. 1324 © 2011 Materials Research Society
DOI: 10.1557/opl.2011.966

Fabrication of High Efficiency Flexible CIGS Solar Cell with ZnO Diffusion Barrier on Stainless Steel Substrate

Bae Dowon[1]†, Kwon Sehan[1], Oh Joonjae[1], Lee Joowon[1], Kim Wookyoung[2]
[1]LG Innotek Components R&D Center, Solar Cell Lab, Ansan, Republic of Korea
[2]Yeungnam Univ., School of Display and Chemical Engineering, Gyeongsan, Republic of Korea
†bdwon@lginnotek.com; choppung@naver.com

ABSTRACT

i-ZnO layers were deposited as diffusion barriers fabricated by RF sputtering on stainless-steel substrates (SUS430, matches with AISI SUS24). It was found that the addition of ZnO layer between stainless-steel substrate and Mo back contact film deplete diffusion of metal ions from substrate and reduce recombination at CIGS layer, as identified by an SIMS depth profile, QE and C-V measurements. With such diffusion barriers, the efficiency, open-circuit voltage, short-circuit current and fill factor of CIGS solar cells all increased, compared to reference cells without diffusion barrier. For the better device performance, Na was supplied during Mo back-contact layer deposition by co-sputtering of the target, including Na-source. Efficiencies of cells were increased with increasing the quantity of Na source. Unlike barrier thickness effect, short circuit current was reduced and open circuit voltage, fill factor were increased with increasing Na-source, and achieved 12.6% efficiency without AR(anti-reflection) coating. The relationship and causality between these results and the Na-doping were analyzed using C-V measurements.

INTRODUCTION

The copper indium gallium diselenide (CIGS) thin-film solar cell recently reached 20.3 percent efficiency [1], setting a new world record for this type of cell. This record is almost same level to efficiency of multicrystalline silicon-based solar cells. CIGS cells use extremely thin layers of semiconductor material applied to a low-cost backing such as glass, flexible metallic foils, high-temperature polymers or stainless steel sheets. Not only efficiency and its flexibility due to the thin absorber, but also a variety of other benefits such as high specific power and aesthetic attractive design, provide a promising path for more affordable solar cells for residential and other uses.

The concept of flexible CIGS based modules on the stainless steel is now accepted as a common idea in the flexible PV industry. Proven high efficiencies of up to η AM1.5=17.5% [2], very high stability against proton and electron radiation [3], potentially light weight, and high specific power as a result, high flexibility seem to fulfill even the desires of demanding customers such as space and transport industries. Moreover, in virtue of its outstanding application property without reference to flections of surface, the enlargement of the application area into BIPV(Building Integrated PV) is also expected. However, despite of these advantages, the most critical subjects are lack of sodium source and the diffusion of detrimental substrate elements, such as Fe, Ni and etc. into the CIGS layer due to the high temperature during CIGS processing, as they deteriorate CIGS cell efficiency [4]. The oxide materials, such as Al_2O_3 and

SiO_2, which can be deposited by PECVD or ALD, are accepted as the general diffusion barrier materials. However, these deposition methods are faced with many disadvantages, such as excessively long process time   and increase of initial investment due to the additional process as shown on table 1. Sodium, which can induce better grain growth of CIGS and device performance, can be supplied by the thermal energy of evaporating process from the glass substrate. Meanwhile, ordinary metal substrate doesn't contain sodium source.

This study we investigate the effect of i-ZnO by RF-sputtering as the barrier layer with regard to the diffusion of Fe or the other metal ions into the CIGS layer, quantity of Na-source supplied during Mo back-contact deposition and its effect on the performance of CIGS solar cells. Main advantages of this methodology over the existing process are simplifying of process steps and reduction of total process time due to the relatively high deposition rate.

Table 1 Comparison of processing time by materials and forming methods

Method	Material	Deposition Rate (nm/min)	Process Time* (300nm)	Remark
ALD	Al_2O_3, SiO_2	1~2	>150 min	
PECVD	Al_2O_3, SiO_2	~20	>15 min	
RF-Sputtering	Al_2O_3, SiO_2	~20	>15 min	
RF-Sputtering	ZnO	~40	> 8 min	Overlap with the TCO process

*Required deposition time under the assumption that continuous in-line process is applied

EXPERIMENT

The investigated steel substrates are listed in Table 1. The stainless steel sheet (SUS24) with a thickness of d=127 μm was selected as substrate because of its lower coefficient of thermal expansion (CTE) compared to nickel chromium steel foil (austenitic).

The steel substrates were cleaned with acetone and alcohol in an ultrasonic bath to remove oil residues from rolling. i-ZnO barrier layer (same material and manufactured by the same equipment as i-ZnO for TCO layer) with thicknesses 100 and 200nm were deposited on substrates by RF sputtering. The Mo back contact with a thickness of about 400 nm was deposited on substrates by DC sputtering. Na precursor layers with various thicknesses were deposited at the bottom of pure Mo back contact by sputtering as sodium doping material to improve device performance (See Fig.1). This concept of flexible CIGS PV can offer several advantages. The first of all we can design factory line simpler and reduce maintenance item due to using same material. And due to the comparatively faster DR(Deposition Rate), we can reduce the tact-time. From the point of view of R&D, it can reduce the initial investment. CIGS absorber layers with a Ga content of [Ga]/([Ga]+[In]) ≈ 0.34 to 0.36 and Cu content of [Cu]/([Ga]+[In])≈0.89 to 0.92 were grown by co-evaporation of the constituent elements in 3-stage process (see ref. [5] for details of the 3-stage process). Solar cells were prepared from the absorber layers by subsequent chemical bath deposition of a CdS buffer layer, RF sputtering of an i-ZnO layer, DC sputtering of a ZnO:Al front contact layer, and electron beam evaporation of Al contact grids. Cell structuring was done mechanically (total cell area without front grid area=0.44 cm^2).

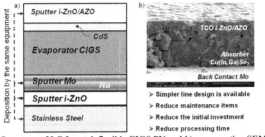

Fig. 1 a) Structure of LG Innotek flexible CIGS PV and b) cross-section SEM image

Solar cells were characterized by standard I-V measurements under AM1.5 equivalent illumination. External quantum efficiency (EQE) measurements were done by measuring the short-circuit current with spectrally resolved monochromatic light. SIMS depth profiling was performed on the CIGS solar cells (with buffer and TCO layers) by a CAMECA IMS 7f system using Cs^+ ions. As the SIMS signal is not quantitative, the Fe signal was calibrated by measuring a CIGS solar cell without diffusion barrier layer. Carrier density change with barrier layers are also measured by C-V measurement.

DISCUSSION

Property Analysis of ZnO diffusion barrier

Diffusion of Fe ions to CIGS absorber from the substrate after using ZnO diffusion barriers as shown in Fig. 2(b) dramatically reduced by increasing thickness. In the case of CIGS fabrication without the barrier layer, the most of Fe ions were kept in the absorber layer.

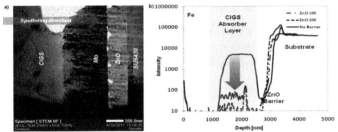

Fig. 2 a) TEM Image of the cross-section and b) SIMS depth profile of Fe ions

From the calculation the relative diffusion rate, it's found that almost 99% of Fe ions were blocked by ZnO layer by increasing the thickness to 200nm (see Table 2). These results are in good agreement with device performances. As shown in Fig 3, we achieved nearly 3% efficiency without barrier layer, but with increasing thickness cell efficiency reached more than 11% (without AR coating) with 200nm ZnO barrier layer. The main factor of efficiency increases in this case was increases of short-circuit currents as shown in Fig 3. In spite of the increase of short-circuit current, fill factor and open-circuit voltage remain still the same after 100nm ZnO. These results indicate that ZnO with 100nm is enough to form normal p-n junction from the point of view of device.

Table 2 Diffusion rate of Fe ions and relative blocking rate

Barrier	no barrier		ZnO 100nm	ZnO 200nm	Relative blocking rate with 200nm ZnO
Zone	Substrate	Absorber	Absorber	Absorber	Absorber
Concentration %	82.90*	5.91	0.07	0.03	99%

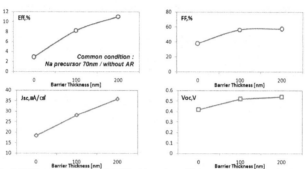

Fig. 3 I-V data of flexible CIGS PV with various thicknesses of ZnO barrier layer

The causality between ZnO barrier and device performance can be inferred from the QE and C-V analysis. As shown in Fig 4(a), QE intensities were dramatically increased at the long-wavelength. Given the result with previous SIMS data, we can easily infer that the metal-ion diffusion was reduced and as the result reduced recombination rate at the absorber layer. Until the 100nm ZnO, reduction of bulk defect led the increase of Jsc, but after 100nm increase of Jsc came from reduction of recombination site at SCR region. These results can be supported by C-V data. As shown in Fig 4(b), Without ZnO barrier it's was found that junction was almost destructed (shunting without regular depletion width), while cells with ZnO barrier showed U-shape graph with depletion width about 0.4 μm (the valley point of U-shape of C-V graph).

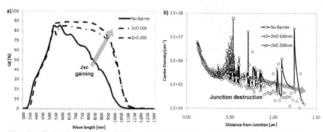

Fig. 4 a) Behavior change of QE and b) C-V by increasing ZnO barrier thickness

118

Na precursor thickness variation effect

From the former test, we achieved more than 11% without AR. To get higher efficiency we additionally have increased the quantity of Na-precursor. Unlike barrier thickness effect, Jsc is reduced and Voc was increased with increasing Na-source. This result agrees with results, which are reported in many papers about Na-effect on Glass-based PV [6]. We gave an attention to the result of 150nm Na-precursor. From the point of view of the main purpose of using Na for CIGS PV, 110nm Na-precursor is enough (Voc increase). But additional Na supply can lead to increase of efficiency by increasing FF.

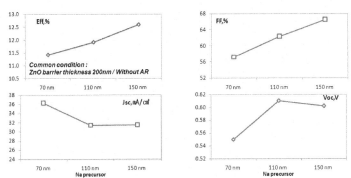

Fig. 5 I-V data of flexible CIGS PV with various thickness of Na precursor

As shown in Fig. (6)a, according to the QE result, Na-doping led to slight Jsc decrease, because Na also can act as defect [7]. But positive effect comes from the increase of carrier density. Carrier density level of flexible CIGS cell has been increased to more than one to the power of 16 levels with the increase of Na-source (see Fig. 6(b)). The increase of carrier density after 110nm, carrier density might be too excessive; as a result, the Voc was reduced. Nevertheless, the continuously increase of FF of the cell with 150nm Na precursor were observed; and as a result 12.6% efficiency was achieved. It's not logical leap that explain this with these analyses, but we can infer that it came from the passivation of grain boundaries by excess-Na [8].

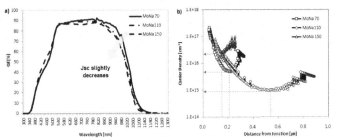

Fig. 6 a) Behavior change of QE and b) C-V data with increasing Na-precursor

119

CONCLUSIONS

It has been shown that ZnO barrier (with thickness 200nm) can reduce 99% of Fe-diffusion to CIGS layer. As a result, defects at bulk CIGS reduced (C-V & QE) and it led to increased Jsc and Eff. Na precursor were used for Na-doping and it led to increased Voc compensating Jsc decrease. Above a certain value of Na thickness, FF still increased even though Voc remained unchanged. ZnO successfully can be substituted for alternative barrier layer without additional equipment investment. There is still room for increasing the efficiency by Na-doping thorough guided carrier density increasing.

ACKNOWLEDGMENTS

The authors would like to thank Dr. Lee Jinwoo from LG Innotek and Park Hyounwook from Yeungnam University for their help in C-V & SIMS analysis throughout this work.

REFERENCES

[1] M. Powalla, W. Wischmanm, F. Kessler, ZSW, Stuttgart, Germany, CIGS Solar Cells with Efficiencies > 20 %: Current Status and New Developments : Oral presentation of the 26th EU PVSEC, Valencia, Spain, Sept. 2010

[2] J.R. Tuttle, A. Szalaj, J. Keane, in: Conference Record of the 28th IEEE Photovoltaic Specialists Conference, Anchorage, AK, USA, Sept. 2000, p. 1042

[3] S.Kawakita, M. Imaizumi, T.Sumita, K. Kushiya, T. Ohshima, M. Yamaguchi, S. Matsuda, S. Yoda, T. Kamiya, in: Proceedings of 3rd World onference on Photovoltaic Energy Conversion, Osaka, Japan, May 11–18. 2003, p. 8PB511.

[4] P. Jackson, P. Grabitz, A. Strohm, G. Bilger, H.-W. Schock, in: Proceedings of the 19th EUPVSEC, Paris, France, 7–11 June 2004, p. 1936

[5] K. Ramanathan, Miguel A. Contreras, Craig L. Perkins, Sally Asher, S Falah, in: Progress in Photovoltaics: Research and Applications Volume 11, Issue 4, pp 225–230, June 2003.

[6] Rudmann, D. da Cunha, A. F. Kaelin, M. Kurdesau, F. Zogg, H. Tiwari, A. N. Bilger, G. : Applied Physics Letters Volume 84, Issue 7, pp 1129-1131, Feb 2004

[7] Jae Ho Yun, Ki Hwan Kim, Min Sik Kim, Byung Tae Ahn, Se Jin Ahn, Jeong Chul Lee, and Kyung Hoon Yoon : Thin Solid Films, Volume 515, Issue 15, pp 5876-5879, 31 May 2007

[8] Uwe Rau, Kurt Taretto, Susanne Siebentritt : Applied Physics A, Volume 96, Number , pp 221-234, Nov 2008.

Mater. Res. Soc. Symp. Proc. Vol. 1324 © 2011 Materials Research Society
DOI: 10.1557/opl.2011.1254

Copper Indium Diselenide thin films using a hybrid method of chemical bath deposition and thermal evaporation

R. Ernesto Ornelas A.[1], Sadasivan Shaji[1,2], Omar Arato[1], David Avellaneda[1], Alan Castillo[1], Tushar Kanti Das Roy[1] and Bindu Krishnan[1,2]

[1]Facultad de Ingeniería Mecánica y Eléctrica, Universidad Autónoma de Nuevo León, San Nicolás de los Garza, Nuevo León, México

[2]CIIDIT- Universidad Autónoma de Nuevo León, Apodaca, Nuevo León, México.

ABSTRACT

Copper indium diselenide (CIS) based solar cells are one among the promising thin film solar cells. Most of the processes reported for the preparation of CIS directly or indirectly involve Se vapor or H_2Se gases which are extremely toxic to health and environment. In this work, we report the preparation of CIS thin films by stacked layers of Glass/In/Se/Cu_2Se and Glass/In/Se/Cu_2Se/Se. For this, first indium (In) thin film was thermally evaporated on glass substrate on which selenium (Se) and copper selenide (Cu_2Se) thin films were deposited sequentially by chemical bath deposition. Selenium thin films were grown from an aqueous solution containing Na_2SeSO_3 and CH_3COOH at room temperature, triple deposition for 7, 7 and 10 min from consecutive baths. Copper selenide thin films were deposited at 35 °C for 1 hour from an aqueous bath containing $CuSO_4$, Na_2SeSO_3 and NH_4OH. Analysis of the X-ray diffraction patterns of the thin films formed at 400 °C from the precursor layer containing extra selenium layer showed the presence of chalcopyrite $CuInSe_2$, without any secondary phase. Morphology of all the samples was analyzed using Scanning Electron Microscopy. Optical band gap was evaluated from the UV-Visible absorption spectra of these films and the values were 1.1 eV and 1 eV respectively for CIS thin films formed at 400 °C from the selenium deficient and selenium rich precursor layers. Electrical characterizations were done using photocurrent measurements. Thus preparation of a $CuInSe_2$ absorber material by a non-toxic selenization process may open up a low cost technique for the fabrication of CIS based solar cells.

INTRODUCTION

$CuInSe_2$ (CIS) belongs to I-III-VI$_2$ group of semiconductor materials and crystallize with tetragonal chalcopyrite structure [1], and is one of the most promising absorbing materials for thin film solar cells, due to high absorption coefficient and direct band gap [2]. CIS thin films have been prepared by several methods such as elemental thermal evaporation [3], binaries compounds thermal evaporation [4], RF sputtering [5], co-sputtering of metallic precursors and selenization [6], pulse laser deposition (PLD) [7], metal organic chemical vapor deposition (MOCVD) [8], electrodeposition [9], chemical bath deposition (CBD) [10] and SILAR [11]. Most of the processes involve heating the precursor layers or a post-deposition treatment in Se vapor/H_2Se gas which are extremely toxic to health and environment. Formation of CIS thin films by heating stacked layers of chemically deposited selenium and vacuum deposited In and Cu was known [12]. Further, use of chemical bath deposited selenium thin films as planar source of selenium to form various binary and ternary selenides thin films were reported [13]. Motivated by these results here we make an effort to explore the advantages of simplicity, low cost and no-toxicity of chemical bath deposition to produce $CuInSe_2$ thin films. In the present work, we report the formation of chalcopyrite $CuInSe_2$ thin films by heating precursor layers of

Glass/In/Se/Cu$_2$Se and Glass/In/Se/Cu$_2$Se/Se at temperatures 350-400 °C, where In layer was thermally evaporated and Se and Cu$_2$Se were deposited by CBD. Optical band gap of the CIS thin films formed were ~ 1 eV. These thin films were photoconductive with dark conductivity value 1.1-2.7 ($\Omega \cdot$cm)$^{-1}$. In our experiment chemical bath deposited Cu$_{2-x}$Se is the precursor for both copper and selenium. However, this selenium is not sufficient to form CuInSe$_2$. Hence, the role of the selenium thin film is to provide the excess selenium for the formation of CuInSe2. CIS thickness and its non-stoichiometry are better controlled in this process than that reported [12].

The preparation process of copper indium diselenide thin films involved heating the precursor layers of indium, selenium and copper selenide layers deposited on glass substrates. The substrates used were glass slides of 25 mm x 75 mm x 1 mm, which were cleaned using neutral detergent followed by washing in water and distilled water, then dried in a hot air flux.

(*i*)*Preparation of precursor layers:* First, indium layer of thickness 100 nm was deposited using thermal evaporation on glass substrate in high vacuum (10^{-5} Torr). Subsequently, these indium coated glass substrates were dipped in an aqueous solution containing 10 ml of sodium selenosulphate (0.1 M), 2 ml acetic acid (25%) and 70 ml distilled water at room temperature. In order to avoid dissolution of indium, selenium deposition was done from new bath consecutively from three baths for 7, 7 and 10 minutes respectively. On the selenium layer, a Cu$_2$Se layer was deposited using CBD at 35 °C for 1 h from an aqueous solution containing 10 ml CuSO$_4$ (0.2 M), 20 ml Na$_2$SeSO$_3$ (0.1 M), 1.5 ml NH$_4$OH and 68.5 ml distilled water (35 °C). Two types of precursor layers were prepared: (a) Glass/In/Se/Cu$_2$Se (labeled as CISA) and (b) Glass/In/Se/Cu$_2$Se/Se (labeled as CISB)

(ii) Heating: These stacked layers of Glass/In/Se/Cu$_2$Se and Glass/In/Se/Cu$_2$Se/Se were annealed at 350 °C and 400 °C in nitrogen atmosphere (0.1 Torr) for 30 minutes. The annealed films were named as CISA350, CISA400, CISB350 and CISB400 depending on the type of precursor layer and the temperature.

(iii) *Characterization*: These thin films were characterized using different techniques. X-ray diffraction (XRD) study was carried out with Cu Kα_1 radiation (λ = 1.5405 Å) diffractometer. Morphology was studied using Scanning Electron Microscope (SEM) (Model JSM-6510 LV JEOL) and compositional analysis was obtained using Energy Dispersive X-Ray Analyzer (EDX) associated with the SEM. Optical properties were studied using optical absorbance spectra recorded by a UV-Vis-NIR spectrometer (Model EPP2000 StellarNet). Dark conductivity and photoconductivity were measured using picoammeter/voltage source (Keithley 6487) interfaced with a computer.

RESULTS AND DISCUSSION

Figure 1 illustrates the XRD patterns of the thin films formed at 350 °C and 400 °C in comparison with that of the as-prepared precursor layers.

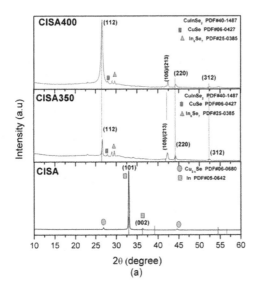

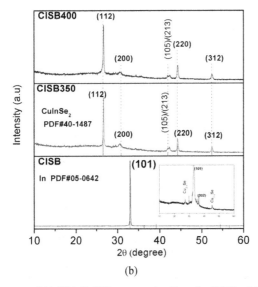

Figure 1 XRD patterns of (a) CISA (b) CISB as prepared and heated at 350°C and 400°C for 30 min. in nitrogen (All are in the same intensity scale). Standard patterns corresponding to pure Indium (PDF #05-0642) and the chalcopyrite CuInSe$_2$ (PDF #40-1487) is included. Zoom in view of CIS to show the peaks corresponding to the entire precursor materials.

In both CISA and CISB, the dominant peak corresponds to (101) plane of metallic indium along with weak reflections due to $Cu_{2-x}Se$ as marked in the figure. In CISA350, indium peak disappears and new peaks are present. The major peaks are identified as reflections from (112), (105),(213) (220) and (312) planes of the chalcopyrite $CuInSe_2$. Also, the presence of CuSe and In_6Se_7 can be seen as marked by two feeble peaks. In CISA400, (112) planes of $CuInSe_2$ grows preferentially. In CISB350 and CISB400, all the peaks are corresponding to chalcopyrite $CuInSe_2$. Reflections due to the binary phases are not seen in these cases. Crystallite size was calculated using Scherrer's formula [14] for CISA400 and CISB400 samples considering the broadening of (112) peak. The values are 18 and 46 nm respectively for CISA400 and CISB400. Thus XRD analysis show that by controlling the selenium content in the precursor layer, formation of single phase chalcopyrite $CuInSe_2$ is achieved.

The morphology of CISA and CISB before and after heating is shown in Figure 2 (a) and (b) respectively. In CISA spherical grains of Cu_2Se (top layer) are distributed uniformly and in CISA350 after heating elongated grains implying synthesis of new compound are observed. This type of elongated features is more prominent in CISA400 than that in CISA350. In the case of CISB, horizontally placed flakes like features of selenium (top layer) are seen. For this precursor after heating at 350 °C and 400 °C these features become flatter and smoother as illustrated in the images of CISB350 and CISB400. EDX measurement results of selenium to metal ratio (Se/Cu+In) for the samples are given in Table 1. In the selenium rich samples (CISBs) this ratio is increased by 75%. Slightly higher value of Se/(Cu+In) for the films formed at 400 °C (CISA400 and CISB400) compared to that in the as-prepared ones (CISA and CISB) may be due to loss of indium as its melting point is low.

Table 1.Compositional analysis by EDX of CIS thin films and electrical conductivity

Samples	Se/(Cu+In)	σ $(\Omega \cdot cm)^{-1}$	$\Delta\sigma$ $(\Omega \cdot cm)^{-1}$
CISA	0.3	-	-
CISA350	0.3	2.1	1.1×10^{-2}
CISA400	0.4	1.9	1.8×10^{-2}
CISB	0.5	-	-
CISB350	0.5	2.7	1.1×10^{-2}
CISB400	0.6	1.1	1.3×10^{-2}

The optical absorbance spectra of CIS thin films formed at 400 °C are shown in figure 3. The spectra were recorded in the wavelength range of 600-1500 nm. From the figure, in CISA350 and CISA400, absorption is higher than that in CISB samples. This may be due to the presence of secondary phases such as CuSe and In_6Se_7 in addition to $CuInSe_2$ phase. However, onset of absorption is well defined in CISB400 near 1200 nm. Since CIS is a direct band gap semiconductor, the absorption coefficient α in an allowed direct transition can be related to the photon energy $h\nu$ by:

$$(\alpha h\nu)^2 = A(h\nu - E_g) \tag{1}$$

The optical band gap (E_g) was determined by extrapolating the straight line portion of $(\alpha h\nu)^2$ vs. $h\nu$ to $\alpha = 0$, where $h\nu$ is the incident photon energy and α is the absorption coefficient determined from the absorbance spectra. These plots are given in figure 3(b) and (c) for the thin films formed at 400 °C. The values are nearly 1.1 and 1 eV respectively for CISA400 and CISB400. These values are in agreement with reported results [15,16]. The decrease in Eg by

~0.1 eV in CISB400 can be attributed to increase in grain size in these thin films due to increase in selenium content. In a study on the effect of selenization of Cu-In precursor using Se vapor [17], it was found that the grain size increased as the selenium content of CIS thin films increased.

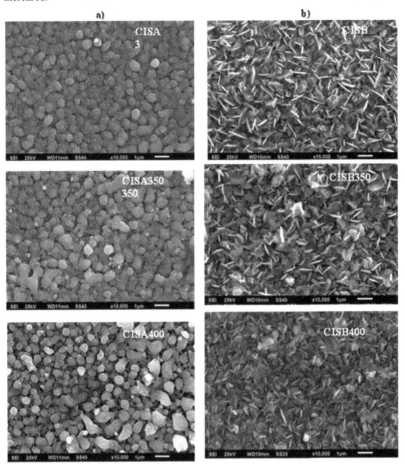

Figure 2: SEM images of (a) CISA (b) CISB before and after heating. All the images are recorded under the same magnification, ×10,000

125

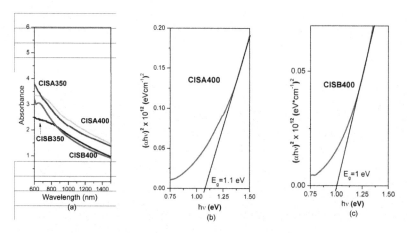

Fig.ure 3: (a) Absorbance spectra of CISA350, CISA400, CISB350 and CISB400 (b) and (c) Evaluation of E_g

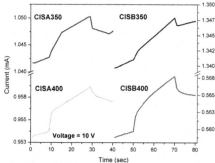

Figure 4: Photocurrent response for CIS thin films

Dark conductivity and photoconductivity of all samples were measured at room temperature. Fig. 4 shows the photocurrent response for CISA350, CISA400, CISB350 and CISB400 samples. The current produced in the sample by applying 10 V was measured under dark and illumination with respect to time. The measurement was done continuously for 40 seconds such that during the first 10 seconds under dark and then for 20 seconds after turn on the light and for last 10 seconds after switching off light. The values of dark conductivity (σ) and increase in conductivity ($\Delta\sigma$) due to illumination are given in Ttable 1. Higher value of conductivity in the thin films formed at 350 °C may be due to the presence of conductive copper selenide phase which is in agreement with the XRD results. In all the graphs, steady increase in current irrespective of the conditions may be due to heating effect from halogen light and high current flow.

CONCLUSIONS

CuInSe$_2$ thin films were prepared by heating precursor layers of Glass/In/Se/Cu$_2$Se and Glass/In/Se/Cu$_2$Se/Se. In this process, indium thin film was grown by thermal evaporation followed by sequential deposition of Se and Cu$_2$Se thin films from chemical baths. CIS thin films possess an optical band gap of approximately 1 eV and they are photoconductive. Thus preparation of a CuInSe$_2$ absorber material by a non-toxic selenization process is achieved. Further investigation is in progress to modify the precursor layer structures and quantify individual layer thickness to obtain the best quality CIS thin films for their applications in solar cells.

ACKNOWLEDGMENTS

The authors are thankful to PAICYT -UANL, Mexico, PROMEP-Mexico and SEP-CONACYT - Mexico (project 105098, 106955) for the financial assistance, One of the authors, Ornelas A. R. Ernesto is grateful to CONACYT-Mexico for providing research Fellowship.

REFERENCES

1. V. G. Lambrecht, Materials Research Bulletin 8 (1973) 1383.

2. A. Luque, S. Hegedus, Handbook of Photovoltaic Science and Engineering, Wiley, New York (2006).

3. K. G. Deepa et al., Solar Energy 83 (2009) 964-968.

4. S. C. Park et al., Solar Energy Materials and Solar Cells 69 (2001) 99-105.

5. J. Müller et al., Thin Solid Films 496 (2006) 364-370.

6. O. Volobujeva et al., Solar Energy Materials and Solar Cells 93 (2009) 11-14.

7. P. Luo et al., Solid State Communications 146 (2008) 57-60.

8. S. H. Yoon et al., Thin Solid Films 515 (2006) 1544-1547.

9. T.-J. Whang et al., Applied Surface Science 257 (2010) 1457-1462.

10. P. P. Hankare et al., Journal of Alloys and Compounds 500 (2010) 78-81.

11. J. Yang et al., Thin Solid Films 517 (2009) 6617-6622.

12. K. Bindu et al., Solar Energy Materials and Solar Cells 79 (2003) 67-79.

13. K. Bindu et al, Journal of The Electrochemical Society 153,(2006) 526-534.

14. B. D. Cullity, R. S. Stock, Elements of X-Ray Diffraction, Third ed., Prentice Hall, New York, 2001, 96-102.

15. R. Caballero, C. Guillen, Solar Energy Materials & Solar Cells 86 (2005) 1–10.

16. S. Agilan, D. Mangalaraj, Sa.K. Narayandass, G. Mohan Rao, Physica B 365 (2005) 93–101.

17. F. Jiang et al, Thin Solid Films 515 (2006) 1950–1955.

III-V and Other Materials

Mater. Res. Soc. Symp. Proc. Vol. 1324 © 2011 Materials Research Society
DOI: 10.1557/opl.2011.838

The two origins of p-type conduction in transparent conducting Ga-doped SnO₂ thin films

Huan-hua Wang [*], Tieying Yang [†], Baoyi Wang, Kurash Ibrahim and Xiaoming Jiang
Institute of High Energy Physics, Chinese Academy of Sciences, Beijing 100049, China

Keywords: tin oxide, net hole conduction, doping, oxygen adsorption, grain boundary
PACS: 81.05.Hd; 68.55.Ln; 68.55.-Nq; 81.15.Cd

* Correspondence author: wanghh@ihep.ac.cn
† Present address: Shanghai Synchrotron Radiation Facility, Shanghai 201204, China

Abstract

The p-type conduction in transparent Ga-doped SnO₂ thin films was realized and its two origins were discerned through comparison experiments associated with growth conditions, Rutherford backscattering spectroscopy and x-ray photoelectron spectroscopy analysis. All the experiment results suggest that the adsorbed oxygen both in the grain boundaries and at the surfaces is another origin of the net hole conduction in the polycrystalline thin films. This mechanism provides a fairy well explanation for the growth temperature dependence of the p-type conductivities of the films. It also offers a useful guide to better the properties of p-type conducting oxide thin films.

INTRODUCTION

p-Type transparent conducting oxide (p-TCO) thin films have very great potentials in applications such as multiple-junction solar cells used in outer space, ultraviolet light emitting diodes, short-wavelength laser devices and so on. Many p-TCOs were successfully synthesized and intensely studied in past decade [1-8]. But by and large speaking, this material family are neither well understood in their fundamental material sciences nor sophisticatedly controlled in their material engineering. The former is embodied by our poor understanding of their defect structures and defect dynamics [9-11]. The latter is demonstrated by the low reproducibility and unsatisfying uniformity of p-TCO thin films and devices. Contrary to the situation of n-type TCOs that have been widely applied in industries like the famous ITO, most p-type TCOs currently are not suitable for practical applications either due to their poor photoelectric properties or due to our inability to solve the problem of their instability or short lifetime. Their instability has to be controlled and their conductivity, hole concentration and mobility must be improved before any industrial application is realized. Therefore, understanding their fundamental microstructure-property relationship is a critical step to improve the fabrication technology and to better the material properties.

Among various p-TCOs, p-type SnO₂ is an excellent candidate for potential industrial applications with many attractive characteristics. It is of simple structure with few components, relatively low growth temperature, low cost, stable thermal and mechanical properties. Compared with p-type ZnO it is more stable in performance. One of its most attractive characteristics is that it can use the already-commercialized n-type SnO₂:F conducting glass as substrate for device applications. Scientists have prepared p-type SnO₂ thin films using Ga [3, 12], In [6], Al [13], Sb [7], Fe [2], Co [14] as acceptor dopants. In these investigations, the acceptor dopants are taken for granted as the only source of net hole conduction. But this is not necessarily the case as our experiments indicated.

We have fabricated p-type Ga-doped SnO_2 (p-SnO_2:Ga) thin films and realized transparent homojunction p-SnO_2:Ga/n-SnO_2:F diodes [12]. In this paper we report the two origins of the p-type conduction of Ga-doped SnO_2 thin films. The oxygen adsorption both in the grain boundaries (GBs) and at the surface of the film was determined to be another source of hole carriers besides the acceptor doping.

Experimental

The thin films were prepared using reactive rf magnetron sputtering. Quartz glass, n-type Si (001), quartz (0001) and sapphire ($10\overline{1}2$) crystals were clamped on the heater as substrates. The growth chamber was pumped to 7.4×10^{-4} Pa and then re-filled with (Ar, O_2) to a dynamically fixed pressure 3.0 Pa with a mixed flow of 20 sccm Ar and 20 sccm O_2. Before each deposition pre-sputtering was maintained for 10 minutes at the set growth temperature. Then the shutter was opened and the growth started. During the sputtering, active oxygen generated by breaking O_2 molecules can suppress the formation of oxygen vacancies. After growth, the samples were naturally cooled down to room temperature.

The microstructures were characterized using x-ray diffraction (XRD) and reflection (XRR) at Beijing Synchrotron Radiation Facility. The electric properties were measured through Hall effect with a van der Pauw geometry. Rutherford backscattering spectroscopy (RBS) and x-ray photoelectron spectroscopy (XPS) was used to measure the component ratios in the film and at the film surface, respectively.

Results and Discussion

The first group of comparison experiments is between the polycrystalline and the monocystalline thin films grown on different substrates under the otherwise same conditions. Fig. 1 shows the XRD profiles of the $Sn_{0.8}Ga_{0.2}O_2$ (nominal stoichiometry in terms of the target) thin films grown at 700°C. Their Hall parameters are listed in Table 1. The films grown on sapphire ($10\overline{1}2$) and quartz (0001) substrates are monocrystalline but electrically insulating, while the films on quartz glass and Si (001) substrates are polycrystalline but p-type conducting. Considering the only difference between the polycrystalline and the monocrystalline thin films, we believe this contrast resulted from the different density of GBs. In the polycrystalline thin films, a large density GBs adsorbed much more oxygen. The adsorbed oxygen in the GBs as well as at the surface depleted the free electrons, and form various chemisorbed oxygen ions [15,16], which caused an upward band bending near the GBs and finally resulted in holes accumulating, as schematically shown in Fig.2. This effect leads to a better bulk p-type conductivity beside the grain boundaries, and even generated quasi-two-dimensional hole gas (2DHG).

It needs to be stated that, as Table 1 shows, the films grown on Si wafers show a lower seeming resistivity and a better seeming stability than that on quartz glass. This is because the SiO_2 surface layer on Si wafer is not continuous [17], which leads to a parallel connection between the film and the Si wafer.

Table 1 The electric properties of typical $Sn_{1-x}Ga_xO_2$ (x=0, 0.2) thin films grown on quartz glass (Q) and Si (001) substrates with the same other sputtering conditions.

Prep. date / Sample	Growth Temp.	Resistivity [$\Omega \cdot$cm]	Carrier dens. [cm^{-3}]	Mobility [cm$^2 \cdot$V$^{-1} \cdot$S^{-1}]	Hall coef. [cm$^3 \cdot$C^{-1}]	Measuring mm/dd/yy
A) 05/17/2009	750°C	3.282	7.886×10^{18}	2.415×10^{-1}	7.925×10^{-1}	06/19/09
Sn$_{0.8}$Ga$_{0.2}$O$_2$/Q		1.178×10^1	1.262×10^{17}	4.203×10^0	4.952×10^1	05/18/10
B) 04/27/2010	700°C	4.493×10^1	6.138×10^{16}	2.266×10^0	1.018×10^2	05/18/10
Sn$_{0.8}$Ga$_{0.2}$O$_2$/Q		6.633×10^0	2.896×10^{18}	3.295×10^{-1}	2.156×10^0	05/21/10
C) 05/10/2010	700°C	2.350×10^1	9.042×10^{16}	2.942	6.912×10^1	05/18/10
SnO$_2$/Q		2.721×10^1	3.374×10^{16}	6.798×10^0	1.850×10^2	05/21/10
D) 05/19/2009	700°C	2.880×10^{-3}	9.514×10^{18}	2.281×10^2	6.569×10^{-1}	06/19/09
Sn$_{0.8}$Ga$_{0.2}$O$_2$/Si		8.410×10^{-3}	1.079×10^{19}	6.887×10^1	5.792×10^{-1}	12/16/09
E) 05/24/2009	700°C	1.278	6.145×10^{16}	7.957×10^1	1.017×10^2	06/19/09
SnO$_2$/Si		1.399	3.769×10^{17}	1.185×10$_1$	1.658×10^1	05/18/10

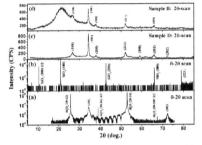

Figure 1. XRD profiles of Sn$_{0.8}$Ga$_{0.2}$O$_2$ thin films grown at 700°C on different substrates: (a) Al$_2$O$_3$ $(10\bar{1}2)$, (b) Quartz (100), (c) Si (001) and (d) quartz glass. The primary x-ray wavelength is λ=1.5448 Å, and its harmonic x-ray diffraction peaks are marked by "$\lambda/2$" behind peak indexes in (a) and (b).

Figure 2. The schematic of oxygen adsorption and the induced depletion region at the grain boundaries (upper) and of the resulted p-type conductivity mechanism (bottom).

The second group of comparison experiments is between the undoped and the Ga-doped SnO$_2$ thin films. Table 1 also lists the Hall parameters of the polycrystalline SnO$_2$ thin films grown on quartz glass and on n-type Si (001) (ρ=4.24 $\Omega \cdot$cm) substrates. The undoped SnO$_2$ thin films exhibited p-type conduction. This undoubtedly indicates there existed excessive oxygen in the films. The undoped film was of a lower p-type conductivity compared with the doped film prepared under the same conditions, implying that the gallium dopant also works as acceptors [12].

133

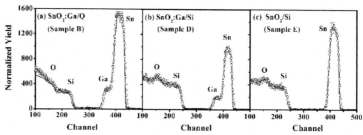

Figure 3. Experimental (open circles) and simulating (solid line) Rutherford backscattering spectra of $Sn_{1-x}Ga_xO_2$ (nominally x=0, 0.2) thin films on quartz glass and on silicon substrates.

Table 2 The fitting results of RBS and the calculated net carrier number per molecular in the doped and undoped thin films supposing the carriers were all delocalized.

Samples	Fitting Results			Calculated net carrier/molec.
	Sublayers	Thickness	Composition ratio	
Sample B:	1. SnGaO	300 Å	Sn 0.170, Ga 0.120, O 0.710	+0.38 hole
$Sn_{0.8}Ga_{0.2}O_2$/Q	2. SnGaO	1070 Å	Sn 0.235, Ga 0.120, O 0.645	−0.01 electron
	3. SnGaO	400 Å	Sn 0.230, Ga 0.120, O 0.670	+0.06 hole
	4. SiO_2	100000Å	Si 1.000, O 2.000	____
Sample D:	1. SnGaO	300 Å	Sn 0.220, Ga 0.100, O 0.680	+0.18 hole
$Sn_{0.8}Ga_{0.2}O_2$/Si	2. SnGaO	770 Å	Sn 0.250, Ga 0.130, O 0.620	−0.15 electron
	3. SnGaO	400 Å	Sn 0.230, Ga 0.120, O 0.670	+0.06 hole
	4. Si	100000Å	Si 1.000	____
Sample E:	1. SnO_2	400 Å	Sn 0.290, O 0.710	+0.26 hole
SnO_2/Si	2. SnO_2	800 Å	Sn 0.370, O 0.630	−0.22 electron
	3. SnO_2	400 Å	Sn 0.360, O 0.640	−0.16 electron
	4. Si	100000Å	Si 1.000	____

The excessive oxygen in the films was directly measured via RBS experiments utilizing a collimated 2.023 MeV He$^+$ beam with a backscattering angle of 165°. Fig.3 shows the experimental data and the fitting curves of the Ga-doped and the undoped SnO_2 thin films. At least a tri-layer model has to be used to achieve good fittings for all the films. The fitting results are listed in Table 2. At the surfaces of both the undoped and doped films, the ratios of O/(Sn+Ga) are always larger than their ideal stoichiometries. So we can safely state that the excessive oxygen at the surface is a hole origin. But only this part of excessive oxygen is not enough to deplete the free electrons generated from oxygen vacancies judging from the case of single-crystal films. So the extra oxygen adsorbed in the GBs of polycrystalline thin film must be a plus to the hole generation by acceptor doping.

The RBS model can be explained according to the structural evolution during the film growth. As the growth started, the strain resulted in small grains [18] and consequently a larger density of GBs adsorbed a large amount of oxygen. As the film grew thicker, the strain was released and the grains developed larger. So the GB density decreased and adsorbed oxygen lessened. In the top layer, the surface adsorbed much oxygen and kept p-type conductive. For the pure SnO_2 thin film on Si, the bottom layer is n-type because the hole density only originating from adsorbed oxygen is overwhelmed by the free electron density coming both from the oxygen vacancies and from the Si substrate.

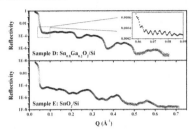

Figure 4. X-ray reflectivity curves of the samples E and F. The inset is the zoom-in part, showing the smallest oscillation cycle of Kiessig fringes.

The automatic formation of tri-layer structure of the film was further cross-checked by XRR measurements. Fig.4 shows the XRR curves of the samples B and E. Apparently multiple layers exist in the samples because at least three oscillation cycles appear in the Kiessig fringes. The left oscillations are abnormal compared with the others, which is due to the heavier absorption of the incident x-rays at small Q values than at large Q values.

As for the adsorbed oxygen at the film surface, an additional verification came from the XPS spectra. The content ratio [O]:[Sn]:[Ga] is 6.5:1.1:1 for the sample E, and the[O]:[Sn] is about 3.3:1 for the sample F, as determined from their peak areas divided by their photoionization cross sections [19]. These rough results indicate that there exists extra oxygen at the film surface.

After taking into account the contribution of adsorbed oxygen to the net holes, the growth temperature dependence of the p-type conductivity of the films can be easily explained. For the $Sn_{0.8}Ga_{0.2}O_2$ thin films, at low temperatures around 150°C, the film is nearly amorphous and thus is insulating due to the Anderson localization. With increasing growth temperatures, the resultant film varied from insulating to n-type conducting, because it crystallized better while much oxygen vacancies still formed in the lattice. As the temperature increased further, the sputtered species reacted with oxygen more quickly, so the oxygen vacancies were suppressed and net holes emerge. When the growth temperature was above 750°C, p-type conductivity degraded because the adsorbed oxygen was desorbed and the grains became larger.

Summary

In conclusion, all the experimental results underlie a point that the p-type conduction in Ga-doped SnO_2 thin film stems from both the gallium dopants and the oxygen adsorbed in the grain boundaries. This point can explain the growth temperature dependence of the p-type conductivity of the samples. Therefore, keeping enough concentrations of gallium dopant and adsorbed oxygen in the film as well as suppressing the n-type defects are the principal guideline to achieve high-quality p-type SnO_2:Ga thin films.

Acknowledgements

This work was financially supported by the National Natural Science Foundation of China (Grant No. 10979057). The authors gratefully acknowledge the support of K.C. Wong Education Foundation, Hong Kong.

References

[1] Q. Mao, Z. Ji, L. Zhao, *Phys. Status. Solidi B* **247**, 299 (2010).

[2] M.-M. Bagheri-Mohagheghi, N. Shahtahamasebi, M.R. Alinejad, A. Youssefi, M. Shokoon-Saremi, *Solid State Sci.* **11**, 233 (2009).

[3] Y. Huang, Z. Ji, C. Chen, *Appl. Surf. Sci.* **253**, 4819 (2007).

[4] M.-L. Liu, L.-B. Wu, F.-Q. Huang, L.-D. Chen, I.-W. Chen, *J. Appl. Phys.* **102**, 116108 (2007)

[5] P.K. Manoj, B. Joseph, V.K. Vaidyan, D.S.D. *Amma, Ceram. Intern.* **33**, 273 (2007).

[6] Z. Ji, L. Zhao, Z. He, Q. Zhou, C. Chen, *Mater. Lett.* **60**, 1387 (2006).

[7] H. Kim, A.Piqué, Appl. Phys. Lett. 84, 218 (2004).

[8] Z. Ji, Z. He, Y. Song, K. Liu, Z. Ye, *J. Cryst. Growth.* 259, 282 (2003).

[9] S. Dutta, S. Chattopadhyay, A. Sarkar, M. Chakrabarti, D. Sanyal, D. Jana, Prog. Mater. Sci. 54, 89 (2009)

[10] S.J. Jokela, M.D. McCluskey, Phys. Rev. Lett. 76, 193201 (2007)

[11] S.B. Zhang, S.-H. We, A. Zunger, Phys. Rev. B 63, 075205 (2001).

[12] T. Yang, X. Qin, H.-H. Wang, Q. Jia, R. Yu, B. Wang, J. Wang, K. Ibrahim, X. Jiang, Q. He, *Thin Solid Films,* **518**, 5542 (2010).

[13] J. Zhao, X.J. Zhao, J.M. Ni, H.Z. Tao, *Acta Mater.* 58, 6243 (2010).

[14] M.-M. Bagheri-Mohagheghi, M. Shokoon-Saremi, *Physica B* **405**, 4205 (2010).

[15] A. Tiurcio-Silver, A. Sánchez-Juárez, *Mat. Sci. Eng. B* **110**, 268 (2004).

[16] T. Sahm, A. Gurlo, N. Barsan, U. Weimar, Sensors and Actuators B **118**, 78 (2006)

[17] T. Suzuki, *J. Appl. Phys.* **88**, 6881 (2000).

[18] J. Sundqvist, J. Lu, M. Ottosson, A. Hårsta, *Thin Solid Film* **514**, 63 (2006).

[19] J.J. Yeh, I. Lindau, *Atomic Data and Nuclear Data Tables* **32**, 1 (1985).

Mater. Res. Soc. Symp. Proc. Vol. 1324 © 2011 Materials Research Society
DOI: 10.1557/opl.2011.1026

DEMUX SiC optical transducers for fluorescent proteins detection

M. Vieira[1,2,3], P. Louro[1,2], M. A. Vieira[1,2], M. Fernandes[1,2], J. Costa[1,2]

[1]Electronics Telecommunication and Computer Dept. ISEL, R. Conselheiro Emídio Navarro, 1949-014 Lisboa, Portugal Tel: +351 21 8317290, Fax: +351 21 8317114, mv@isel.ipl.pt
[2] CTS-UNINOVA, Quinta da Torre, Monte da Caparica, 2829-516, Caparica, Portugal.
[3] DEE-FCT-UNL, Quinta da Torre, Monte da Caparica, 2829-516, Caparica, Portugal

ABSTRACT

This paper presents results on the optimization of multilayered a-SiC:H heterostructures that can be used as an optical transducer for fluorescent proteins detection. Stacked structures composed by p-i-n based a-SiC:H cells are used as wavelength selective devices, in the visible range. The transfer characteristics of the transducers are studied both theoretically and experimentally under several wavelength illuminations corresponding to different fluorophores and tested for a proper fine tuning in the violet, cyan and yellow wavelengths. The devices were characterized through spectral response measurements under different electrical and optical bias conditions and excitation frequencies. Results show that the output waveform is balanced by the wavelength and frequency of each input fluorescent signal, keeping the memory of the wavelength and intensity of the incoming optical carriers. To selectively recover a single wavelength a specific voltage or optical bias is applied.

INTRODUCTION

Over the past decade, fluorescent proteins have launched a new era in cell biology by enabling investigators to apply routine molecular cloning methods, fusing these optical probes to a wide variety of protein, in order to monitor cellular processes in living systems using fluorescence microscopy and related methodology. The spectrum of applications for fluorescent proteins ranges from reporters of transcriptional regulation and targeted markers to fusion proteins designed to monitor motility and dynamics. These probes have also opened the door to creating biosensors for numerous intracellular phenomena. By applying selected promoters and targeting signals, fluorescent protein biosensors can be introduced into an intact organism and directed to a host of specific tissues, cell types, as well as sub cellular compartments to enable a unique focus on monitoring a variety of physiological processes [1, 2]. There is great demand of autonomous systems able of rapid, highly sensitive, in situ characterization and quantification of chemical and biological species and in-vivo monitoring. For this purpose, biosensors must have remarkable sensitivity, specificity and efficiency due to the selective evolution of molecular mechanisms. We report results on the optimization of a-SiC:H heterostructures that can be used as optical transducer for fluorescent proteins detection.

EXPERIMENTAL DETAILS

Device optimization and operation

Voltage and optical bias controlled devices, with front and back indium tin oxide transparent contacts, were produced by PECVD at 13.56 MHz radio frequency and tested for a proper fine tuning of the visible spectrum. The active device consists of a p-i'(a-SiC:H)-n / p-i(a-Si:H)-n heterostructure with low conductivity doped layers ($<10^{-7}\Omega^{-1}cm^{-1}$) . The thicknesses and optical gap of the thin i'- (200nm; 2.1 eV) and thick i- (1000nm; 1.8eV) layers are optimized for light absorption in the blue and red ranges, respectively [3]. Thus, photo generation occurs firstly in the a-SiC:H absorber, and the remaining non -absorbed light goes through the a-Si:H layer. As result, both front and back diodes act as optical filters confining, respectively, the blue and the red optical carriers, while the green ones are absorbed across both [4]. The devices are optimized for the detection of the green fluorescent protein (egFP) and its color-shifted genetic derivatives developed by engineering specific mutations in the original GFP nucleotide sequence, spanning the cyan (CFP), yellow (YFP), red (dTomato) regions of the visible spectrum. In Figure 1 the normalized spectral emission of some fluorescent proteins are displayed as well as the device configuration and light penetration depths.

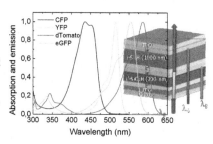

Figure 1 Device configuration (insert) and normalized spectral proteins' emissions.

Optical and voltage controlled light filtering

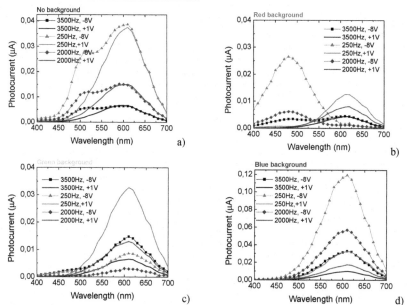

a)

b)

c)

d)

Figure 2 Spectral photocurrent under positive (+1V) and negative (-8V) applied voltage and different frequencies without additional optical bias (a) or under steady state red (b), green (c) and blue (d) irradiations (background).

In order to analyze the spectral sensitivity of the device under different excitation frequencies, light bias and applied voltages, spectral response measurements without and with steady state applied optical bias and current-voltage characteristics were performed. In Figure 2 the spectral photocurrent at different frequencies is displayed under positive and negative applied voltages: a) without additional optical bias (φ=0) or under steady state irradiation: b) red (λ=624 nm); c) green (λ=526 nm) and d) blue (λ=470 nm) backgrounds. In Figure 3a it is displayed the measured photocurrents at 470nm, 526nm and 624 nm without (lines) and under red irradiation (symbols) as a function of the applied voltage. In Figure 3b the spectral gain, defined as the ratio between the photocurrent, under red, yellow, green, blue and violet steady state illumination and without it, is plotted at -8V.

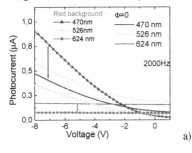 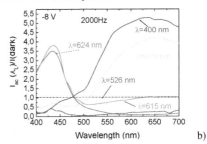

a)   b)

Figure 3 a) Photocurrents at 470nm, 526nm and 624 nm without (lines) and under red irradiation (symbols) as a function of the applied voltage. b) Ratio between the photocurrent, under red, yellow, green, blue and violet steady state illumination and without it, at -8V.

Results from Figure 2a and Figure 3a show that without applied optical bias, in the long wavelength range (> 600 nm), the spectral response is independent of the applied bias. An opposite behavior is found in the short wavelength range as, in this part of the spectrum, the collection strongly increases with the reverse bias. Under steady state optical bias and negative applied voltage, the blue background (Figure 2d) enhances the light-to-dark sensitivity in the long wavelength range and quenches it in the short wavelength range. The red bias (Figure 2b, Figure 3a) has an opposite behavior; it reduces the ratio in the red/green wavelength range, mainly under reverse bias, and amplifies it in the blue one. Under green background no self amplification was detected (Figure 2c).

The spectral sensitivity depends also on the frequency (Figure 2). As the frequency increases the spectral photocurrent decreases suggesting unbalanced capacitive effects between both front and back diodes.

When an external electrical bias (positive or negative) is applied to a double pin structure, its main influence is in the field distribution within the less photo excited sub-cell. The front cell, under red irradiation; the back cell, under blue light, and both, under green steady state illumination. The field under illumination is lowered in the most absorbing cell (self forward bias effect) while the less absorbing reacts by assuming a reverse bias configuration (self reverse bias effect) [5].The front diode, based on a-SiC:H heterostructure, cuts the wavelengths higher than 550nm (absorbed in the back diode) while the back one, based on a-Si:H, cuts the ones lower than 500nm (absorbed in the front diode). When the electrical field increases locally (lest excited cell) the collection is enhanced and the gain is higher than one. If the field is reduced (most excited cell) the collection is reduced and the gain is lower than one. Data from Figure 3b confirms that the sensor is a wavelength current-controlled device that makes use of changes in the wavelength of the optical bias to control the power delivered to a load, acting also as an optical amplifier. Its gain depends on the voltage and on the background wavelength that both control the electrical field profile across the device. Those bias controlled optical nonlinearities make the transducer attractive for fluorescence based proteins sensing. It can detect their emission spectrum and intensities providing a relatively flat partial gain spectrum which makes possible to use multiple wavelengths.

OPTICAL ENCODED DATA STREAM.

A chromatic time dependent wavelength combination (4000 bps, 2000Hz) of R (λ_R=624 nm), G (λ_G=526 nm) and B (λ_B=470 nm) pulsed input channels with different bit sequences, was used to generate a multiplexed signal in the device. The output photocurrents, under positive (+1V) and negative (-8V) voltages with (color lines) and without (black lines) background is displayed in Figure 4. The bit sequences are shown at the top of the figures.

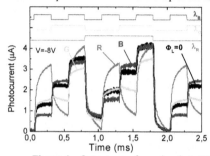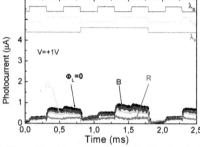

Figure 4   Output waveform signal at -8V and +1V; without (φ_L=0) and with (R, G, B) optical bias. The bit sequences are shown at the top of the figures.

As in Figures 2 and 3, even under transient input signals (the input channels), the background wavelength and the applied voltage control the output signal in a mode that induces a nonlinear wavelength bias dependent gain (Figure 3). This nonlinearity, due to the asymmetrical light penetration of the input channels across the device together with the modification on the electrical field profile due to the electrical and optical bias, allows tuning an input channel without demultiplexing the optical encoded data stream.

To recover the transmitted information (4000bps, 8 bit per wavelength channel) the output waveforms under red irradiation and without it were used as displayed in Figure 5. Both multiplexed signals, during a complete cycle (T), were divided into eight time slots (Δ=250 µs) corresponding to one bit where the independent optical signals can be ON (1) or OFF (0). In Figure 5 all the possible combination of the three input channels are present, so, the waveform of the output without optical bias is an 8-level encoding (2^3) to which it corresponds 8 different photocurrent thresholds. Taking into account Figure 3b, under red background the red channel

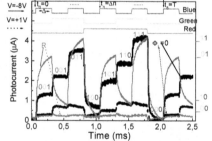

Figure 5   Output waveform signal at -8V and +1 V; without (φ_L=0) and with (R) optical bias.

is strongly quenched and the blue enhanced. So, the output waveform becomes a main 4-level encoding (2^2, right axis). Here, the higher level corresponds to both blue and green ON (_11) and the lower to the absence of both (_00). The other two intermediate levels are ascribed, the upper level to the ON state of the blue (_01) channel and the lower to the green channel ON (_10). To decode the red channel, the same time slots without and with red background has to be compared, being higher the ones with the red channel ON. Using this simple algorithm the independent red, green and blue bit sequences were decoded as: B[10101010], G[01100110] and R[00011110].

Another way to decode the RGB information is using the voltage bias controlled sensitivity of the device (Figure 3a). Under positive bias the device has no sensitivity to the low wavelengths (blue channel) while under negative bias its sensitivity is strongly enhanced. So the multiplex signal at +1V is a 4-level encoding where the higher level corresponds to both red and green channels ON (11_)and the lower to the absence of both (00_), the intermediate levels are due, the upper level tothe ON state of the green (01_) channel and the lower to the red channel ON (10_). To retrieve the blue information equivalent time slots bits under positive bias have to be compared under negative bias and the higher levels ascribed to the blue channel ON.

FRET APPROACH

Many proteins have benefited from the process of selective evolution to become very sensitive to specific molecular species such that in their presence conformational changes and binding events take place. One possible approach to detect such events is Fluorescence Resonance Energy Transfer (FRET), a mechanism by which the fluorescence wavelength changes if two labeled molecules come close enough [6].

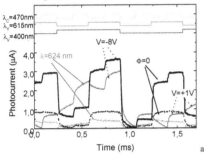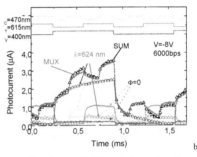

Figure 6 a) Multiplexed signals with and without red background. b) Input channels with red irradiation and without it at -8V. On the top, the optical signal used to transmit the information.

To simulate the excitation light and the FRET pairs (CFP, YFP) three modulated (6000bps) monochromatic beams: Violet (V; λ_V =400nm); Cyan (C; λ_C =470nm); Yellow (Y; λ_Y =615 nm) with different bit rates and their polychromatic combinations illuminated separately the device from the glass side, and the photocurrent was measured with and without the red background under negative (-8V) and positive (+1V) voltages. Figure 6a displays the combined signals due to transmission of the three independent sequences, each one assigned to one color channel without (black line) and under red background irradiation (color lines). The reference level was assumed to be the signal when all the input channels were OFF. At the top of the figure, the individual optical signals are displayed to guide the eyes in relation to the different ON-OFF states. The independent bit sequences (8 bit per wavelength channel) were chosen in order to sample all the possible chromatic mixtures for a pulse rate of 6000bps. In Figure 6b the input channels with red irradiation and without it are displayed at -8V. For comparison the sum of the individual channels at -8V (SUM) and the multiplexed signal (MUX), adjusted to their minimum values, are displayed.

A good fit was obtained showing the independence of the three input channels. Results show that under negative voltage with and without optical bias the waveform of the output signals are always 8-level encoding (2^3). Under positive bias or red irradiation the levels are reduced to 2-level encoding (2^1) due to the lower sensitivity of the device to the cyan and violet (Figures 2 and 3) allowing the immediate decoding of the yellow channel. Once the yellow channel decoded the other two can be obtained from the MUX signal at -8V taking into account their amplitude dependence with the applied bias (Figure 5). To recover the CFP and YFP emission intensities, red optical bias was used. Under red irradiation the yellow channel is quenched, the blue enhanced and the violet strongly amplified (Figure 3b and Figure 6b). So the highest four levels in the MUX signal (Figure 6b) corresponds to the presence of the excitation light (the violet) and the lowest four levels to its lack. The yellow channel is decoded under positive bias (2-level encoding). By subtracting the yellow coding to the MUX signal a 4 level encoding is obtained. Here the highest amplitude corresponds to the presence of both violet and cyan channels ON, the lowest to their lack and the two intermediated levels respectively to the presence of the violet or of the cyan ON. By using this simple algorithm the emission spectra of the CFP and YFP is recovered, in real time, and its ratio can be correlated with the distance between the fluorophores.

CONCLUSIONS

An optical transducer for fluorescent protein detection in multicolor scenarios was presented. Different emission colors of fluorescent proteins were spectrally distinguished from their simultaneous emissions. Results show that the output waveform is balanced by the wavelength and frequency of each input fluorescent signal, keeping the memory of the wavelength and intensity of the incoming optical carriers. To selectively recover only one wavelength, a specific voltage or optical bias has to be applied. The performance of the algorithm is assessed either by using light or positive bias.

ACKNOWLEDGEMENTS

This work was supported by FCT (CTS multi annual funding) through the PIDDAC Program funds and PTDC/EEA-ELC/111854/2009.

REFERENCES

[1] D.A LaVan, Terry McGuire, Robert Langer, Nat. Biotechnol. 21 (10), 2003, 1184–1191.
[2] D. Grace, Medical Product Manufacturing News, 12, 2008, 22–23.
[3] M. Vieira, A. Fantoni, M. Fernandes, P. Louro, G. Lavareda, C. N. Carvalho, Journal of Nanoscience and Nanotechnology, Vol. 9, , Number 7, July 2009 , pp. 4022-4027(6).
[4] P. Louro, M. Vieira, Yu. Vygranenko, A. Fantoni, M. Fernandes, G. Lavareda, N. Carvalho, Mat. Res. Soc. Symp. Proc., 989 (2007) A12.04.
[5] M. Vieira, A. Fantoni, P. Louro, M. Fernandes, R. Schwarz, G. Lavareda, and C. N. Carvalho, Vacuum, Vol. 82, Issue 12, 8 August 2008, pp: 1512-1516.
[6] Jovin, T.M. and Arndt-Jovin, D.J. FRET microscopy: Digital imaging of fluorescence resonance energy transfer. Application in cell biology. In Cell Structure and Function by Microspectrofluometry, E. Kohen, J. G. Hirschberg and J. S. Ploem. London: Academic Press, 1989. pp. 99-117.

Mater. Res. Soc. Symp. Proc. Vol. 1324 © 2011 Materials Research Society
DOI: 10.1557/opl.2011.841

Transmission electron microscopy of misfit dislocation and strain relaxation in lattice mismatched III-V heterostructures versus substrate surface treatment

Y. Wang,[1] P. Ruterana,[1] L. Desplanque,[2] S. El Kazzi,[2] and X. Wallart[2]

[1]CIMAP UMR 6252 CNRS-ENSICAEN-CEA-UCBN, 6, Boulevard du Maréchal Juin, 14050 Caen Cedex, France
[2]Institut d'Electronique, de Microélectronique et de Nanotechnologie, UMR-CNRS 8520, BP 60069, 59652 Villeneuve d'Ascq Cedex, France

ABSTRACT

High resolution transmission electron microscopy in combination with geometric phase analysis is used to investigate the interface misfit dislocations, strain relaxation, and dislocation core behavior versus the surface treatment of the GaAs for the heteroepitaxial growth of GaSb. It is pointed out that Sb-rich growth initiation promotes the formation of a high quality network of Lomer misfit dislocations that are more efficient for strain relaxation.

INTRODUCTION

In the last decade, growth of antimony-based III-V semiconductors have been attracting much attention for potential applications in high-speed and low-power electronic and optoelectronic devices, due to their wide range bandgaps, small electron effective mass and high electron mobility [1-3]. Unfortunately, the strain and high density of defects due to the large mismatch between the III-Sb epitaxial layers and substrate (e.g. GaAs, GaP, Si) has until have an obstacle for both the electrical and optical properties of the devices. Because of the large mismatch (7.8%, in the case of GaSb/GaAs), the critical thickness is expected to be within the range of few monolayers (MLs) and subsequently the misfit dislocations are generated at the interface to release the misfit strain. The growth processes of hetero-structures have been largely investigated, and both 90° Lomer and 60° misfit dislocation were reported form at the interface [4-6]. For the epitaxial growth of semiconductor, it is known that the surface treatment plays a critical role in the growth process (growth model, surface morphology, stoichiometry …) of the epitaxial layer [7].

In the present work we report on a study by means of conventional TEM and high-resolution TEM (HRTEM) combined with geometric phase analysis method (GPA) on the interface misfit dislocations and strain relief in the GaSb epitaxial grown on (001) GaAs with Ga-rich or Sb-rich surface treatment. This report is focused on the possible effects of the surface treatment of (001) GaAs on the arrangement of the interface misfit dislocations and strain relaxation.

EXPERIMENT

The investigated GaSb layers were grown on GaAs (001) ±0.5° semi-insulating substrates by Molecular Beam Epitaxy in a 3-inch Riber Compact 21TM reactor with a base pressure better than 1×10^{-10} Torr. After de-oxidation at 625°C under an As flux, a 500 nm GaAs layer was first grown at 580°C to smooth the surface. Then the As valve was closed and the sample temperature was decreased to 510°C under Sb_2 flux for sample B and without any flux for sample A. For sample A, 1 ML Ga was deposited just before growing the GaSb layer. The growth rate was 0.7 ML/s for the antimonide layers; the growth process was monitored by in situ reflection high-energy electron diffraction (RHEED). The Sb_2 exposure during cooling of the GaAs buffer leads to a (2x8) RHEED pattern whereas for sample A the starting reconstruction was a (2x4) one. During the initial steps, the RHEED pattern exhibited rapidly a 3D transition indicative of a Volmer Weber growth mode when GaSb growth began. After a few nanometers, we recovered a 2D RHEED pattern with a 1x3 surface reconstruction. The GaSb layer thickness was 600 nm.

Plan-view and cross-sectional samples were prepared for TEM and HRTEM investigations. For the cross-sectional samples, slices of 2x5 mm^2 in size were cut from the substrate side along the [110] and [1-10] directions. Two of the slices were glued face to face and packed in a copper tube of 3 mm in diameter with the epoxy glue, then the tube were cut in to disk of about 800 µm in thickness. These disks were then mechanically polished and dimpled from both sides until the thickness of the central area was about 10 µm. The final thinning was performed by argon ion milling at -150°C in order to minimize ion beam damage. For the plan-view sample 3x3 mm^2 slabs were polished, dimpled and ion milled only from the substrate side. The conventional TEM and HRTEM analysis were carried out on two JEOL microscopes: 2010 LaB_6 and 2010 FEG both operated at 200 kV, respectively.

RESULTS AND DISCUSSION

In III-V cubic compounds, as the dislocations have a/2 <110> Burgers vectors some of them will be out of contrast in observations carried out in cross sections along <110> type zone axis. Therefore, in order to determine the threading dislocation density, we have carried out plan-view observations along the [001] zone axis. Figure 1 shows plan-view TEM micrographs of sample A and B: the threading dislocations appear as dark dots on the density background, they are marked by arrows.

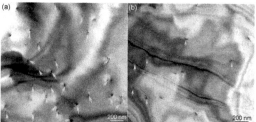

Figure 1: Plan-view TEM images of sample A and B, recorded close to the [001] reflection.
The averaged threading dislocations estimated from several images are 7.5×10^8 and 2.2×10^8 threading dislocations/cm^2 for samples A and B, respectively. Obveriously, with Sb-rich surface treatment the threading dislocation density decreased. In order to check how the surface treatment affect the misfit dislocatiosns and strain relaxation, we have carried out a detailed HRTEM analysis of the two samples. Figure 2 shows HRTEM images of the (110) interface between the GaSb epitaxial layer and GaAs substrate for sample A (a) and B (b). In these images, the positions of the interface dislocations have been marked by the additional {111}

lattice planes (inclined arrows). As the lattice constant of Gasb is larger than that of the GaAs substrate, the extra half planes of the misfit dislocations are observed in the GaAs substrate. Lomer dislocations and closely spaced 60° dislocation pairs (or dissociated Lomer dislocation) can be seen along the interface. These defects are the major interfacial defects which accommodate the misfit strain. As showing in Figure 3, in sample A, the misfit dislocations are essentially made of 60° dislocation pairs (or dissociated Lomer dislocation) of which one in the epitaxial layer, the second in the substrate. In contrast, all the extra {111} planes terminate at the interface and Lomer dislocation are the major misfit dislocation at the interface in sample B. Based on several images, we have determined the mean spacing of the Lomer dislocation are 5.67 nm and 5.46 nm for sample A and B, respectively. Evidently, for sample B, the average distance of the interface dislocations almost coincides with the theoretical value (5.51 nm) for Gasb/GaAs heterostructures, indicative of a higher strain relaxation state in the epitaxial layer.

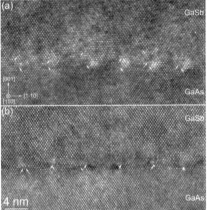 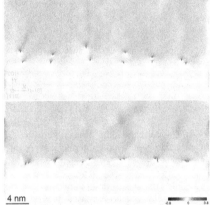

Figure 2: Cross-sectional HRTEM images of the Gasb on (001) GaAs with Ga-rich (a) and Sb-rich (b) surface observed along [110] orientation. The inclined arrows indicate the extra {111} planes close to the interface.

Figure 3: Strain ε_{xx} components corresponding to the Fig. 2 HRTEM images. In sample A all the interface misfit dislocation cores are split in two. In sample B, the cores of the Lomer dislocations are compact.

To investigate the local strain distribution in these samples, we apply the GPA [8-10] of the HRTEM images. The GPA method relies on the evaluation and interpretation of the geometric phase component $P_g(r)$ by performing a Fourier transform on a HRTEM image. For perfect crystals, the phase of a Bragg-reflection, described by the reciprocal space vector g, is constant across the image. However, for a distorted lattice, small deformation can be seen as local lateral shifts of the lattice fringes and consequently as small changes in the phase corresponding to g. The phase $P_g(r)$ determined by GPA is related to the displacement field by the expression

$$P_g(r) = -2g \cdot u(r)$$

From the local lattice displacements, the two-dimensional strain maps can then be calculated as

$$\varepsilon_{xx} = \frac{\partial u_x}{\partial x}, \ \varepsilon_{xy} = \frac{1}{2}\left(\frac{\partial u_x}{\partial y} + \frac{\partial u_y}{\partial x}\right), \ \varepsilon_{yx} = \frac{1}{2}\left(\frac{\partial u_y}{\partial x} + \frac{\partial u_x}{\partial y}\right), \ \varepsilon_{yy} = \frac{\partial u_y}{\partial y},$$

Figure 3 (a) and (b) shows the ε_{xx} component of the strain field (x axis along the [1-10]) derived from Figure 2. On these images the dislocation cores are easily located, as they corresponding to the areas where the strain is maximal. Besides the strain distribution, the distance of the split cores in sample A is larger than that in sample B. To quantify the strain relaxation state, one can project the ε_{xx} on the growth direction (y axis). As show in figure 4, the value of ε_{xx} in sample B is larger than sample A indicative of a better strain relaxation in sample B. We have observed a similar behavior in the GaSb/GaP system at the initial growth stage [11]. Moreover, the abrupt change in the intensity at the interface region reveals that the thickness of the dislocation cores region is 2.93 nm and 0.95 nm for sample A and B, respectively. This clean decrease in the thickness of the dislocation cores region indicates a sharp interface between the GaSb epitaxial layer and GaAs substrate with Sb-rich surface treatment.

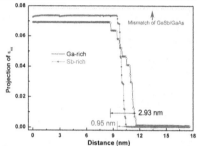

Figure 4: The corresponding projection of the ε_{xx} images on the growth direction, the vertical part of each curve shows the interfacial layer thickness and the max height corresponds to the relaxation level.

Besides the strain distribution, the fine structure of the dislocation core has also been analyzed using the method proposed by Kret et al. [12,13] considering the continuum theory of the dislocation. This approach provides a quantitative determination of the Burgers vector of the misfit dislocation and direct insight on their position at the interface. Figure 5 shows the α_{13} components of the tensorial distribution of the dislocation cores derived from the HRTEM images. In these images, the dislocation core density tensor takes non-zero values only close to the dislocation core. Integrating the α_{13} over the dislocation core region, we can obtain the in-plane components of the Burgers vector as shown in the figure 5. It can be seen that the calculated Burgers Vector are very close to the theoretical value for the Lomer dislocation (b=a/2 [1-10] =4.00 Å). Moreover, integrating the dislocation density peaks separately, we obtain two Burgers vectors corresponding to in plane components of two 60° dislocations. This agrees with the fact that Lomer dislocation is formed by the reaction of two 60° dislocation pairs [14]. As determined in many areas, the distance between the dissociated cores for the misfit dislocations have been summarized in figure 6. As can be seen, the dislocation cores of sample B are more localized. For comparison of the two samples, we arbitrarily define a pure Lomer dislocation as characterized by two additional intersecting lattice planes and a splitting of 60° dislocation cores less than 1.5 nm. We then have 64.7 % and 93.5 % of Lomer dislocations for sample A and B, respectively. The increase in the Lomer dislocation fraction indicates that the Sb-rich GaAs surface treatment promotes the formation of Lomer misfit dislocations, which leads to a better strain relaxation in the epitaxial layer.

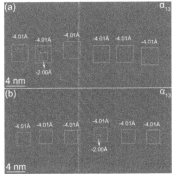

Figure 5: The α_{13} component of the dislocation distribution tensor field, the quantified Burgers vector have been indicated in the images.

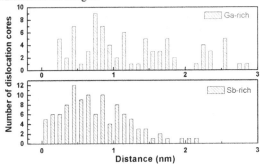

Figure 6: Statistical distribution of the distance between the cores of misfit dissociated dislocations in the two samples.

CONCLUSIONS

The misfit dislocation and strain relaxation of the highly mismatched GaSb on (001) GaAs with Ga-rich and Sb-rich surface treatment has been studied. As shown above, the GaSb epitaxial layer grown on GaAs substrate subsequent to a Sb-rich surface treatment exhibits a lower threading dislocation density and better strain relaxation than on a Ga-rich surface. This HRTEM and GPA analysis has shown that the strain relaxation state in the epitaxial layer is related to the formed interface misfit dislocations; Sb-rich (001) surface enrichment is found to promote the formation of Lomer misfit dislocations which are more efficient for strain relaxation.

ACKNOWLEDGMENTS

This work is supported by the national research agency under project MOS35, contract No.: ANR-08-NANO-022.

REFERENCES

1. P.S.Dutta, H.L.Bhat, and V.Kumar, J.Appl.Phys. **81**, 5821 (1997).
2. R.M.Biefeld, Mater. Sci. Eng. R **36**, 105 (2002).
3. B.R.Bennett, R.Magno, J.B.Boos, W. Kruppa, and M.G.Ancona, Solid-State Electron. **49**, 1875-1895 (2005).
4. A.Vila, A.Cornet, J.R.Morante, P.Ruterana, M.Loubradou, R.Bonnet, Y. Gonzalez, and L.Gonzalez, Philos.Mag.A **71**, 85 (1993).
5. A.Vila, A.Cornet, J.R.Morante, P.Ruterana, M.Loudradou and R.Bonnet, J. Appl. Phys. **79**, 676 (1996).
6. W.Qian, M.Skoronski, R.Kaspi, M.Degraef, and V.P.Dravid, J. Appl. Phys. **81**, 7268 (1997).
7. B.A.Joyce, and D.D.Vvedensky, Mater. Sci. Eng. R **46**, 127 (2004).
8. M.J.Hytch, E.Snoek, and R.Kilaas, Ultramicroscopy **74**, 131(1998).
9. S.Kret, P.Ruterana, A.Rosenauer, and D.Gerthsen, Phys. Stat. Sol. (b) **227**, 247 (2001).
10. Y.Wang, P.Ruterana, L.Desplanque, S.EI Kazzi, and X.Wallart, J.Appl.Phys. **109**, 023509 (2011).
11. S.EI Kazzi, L.Desplanque, C.Coinon, Y.Wang, P.Ruterana, and X.Wallart, Appl.Phys.Lett. **97**, 192111 (2010).
12. S.Kret, P.Dluzewski, P.Dluzewski, and E.Sobczak J.Phys.:Condens.Mater **12**, 10313 (2000).
13. S.Kret, P.Ruterana, and G.Nouet, J.Phys.:Condens.Mater **12**, 10249 (2000).
14. J.Narayan, and S.Oktyabrsky, J.Appl.Phys. **92**, 7122 (2002).

Mater. Res. Soc. Symp. Proc. Vol. 1324 © 2011 Materials Research Society
DOI: 10.1557/opl.2011.843

Structure and photocatalytic properties of $Bi_{25}FeO_{40}$ crystallites derived from the PEG assisted sol-gel methods

Shundong Bu, Dengrong Cai, Jianmin Li, Shengwen Yu, Dengren Jin and Jinrong Cheng
School of Material Science and Engineering, Shanghai University, Shanghai, China

ABSTRACT

Sillenite $Bi_{25}FeO_{40}$ crystallites have been fabricated via a sol-gel approach. X-ray diffraction results show that single-phase $Bi_{25}FeO_{40}$ can be synthesized at the annealing temperature of 600 °C with the help of PEG additive. The amount of additives and the annealing temperature has great effects on the formation of phase pure $Bi_{25}FeO_{40}$ crystallites. The morphologies of $Bi_{25}FeO_{40}$ crystallites were observed by SEM techniques. UV-vis diffuse reflectance spectroscopy indicated the good visible light absorption of $Bi_{25}FeO_{40}$ crystallites. The photo-catalytic activity of $Bi_{25}FeO_{40}$ powders was evaluated by the degradation of methyl orange solution assisted by H_2O_2 under UV-Vis light and Vis-only light irradiation, which suggested that $Bi_{25}FeO_{40}$ crystallites are potential photocatalytic materials.

INTRODUCTION

Akira Fujishima discovered that water was photocatalytic decomposed at a TiO_2 electrode in 1972 [1]. Since then, semiconductor photocatalyst like TiO_2 is widely used to degrade organic pollutant and produce hydrogen as clean energies. Nowadays, TiO_2 is the most interesting catalytic materials due to its high activity , chemical stability and non-toxic. However, the band gap of TiO_2 is 3.2 eV so that it can only respond under ultraviolet irradiation, which means plenty of sunlight including visible light and infrared ray was useless for it. Therefore, many new semiconductor photocatalytic materials with low energy gap appears.

Sillenite material is one of the novel photocatalysts. Due to its unique BiO heptacoordination in the crystal structure [2], sillenite compounds , with the general formula $Bi_{12}MO_{20}$ (M=Ge, Ti, Ga, Fe, Bi, V, etc.), has been reported the high photocatalytic property and photo-conductivity, such as $Bi_{12}SiO_{20}$ [3], $Bi_{12}GeO_{20}$ [4], $Bi_{12}TiO_{20}$ [5] and $Bi_{24}Ga_2O_{39}$ [6] . It is reported [7] that $Bi_{25}FeO_{40}$ also has the structure of sillentite with $M=Bi_{0.5}Fe_{0.5}$. The umbrella-like $Bi^{3+}O_3$ groups [8] in $Bi_{25}FeO_{40}$ crystal is also considered to eliminate the recombination of electron–hole pairs and its photocatalyst behavior is worth looking forward to.

In this paper, single-phase $Bi_{25}FeO_{40}$ crystallites were fabricated by PEG-assisted sol-gel method. The morphology and the photocatalytic activity of $Bi_{25}FeO_{40}$ powder is also reported.

EXPERIMENT

The sillenite $Bi_{25}FeO_{40}$ were synthesized by sol-gel method. The chemicals used in the experiment is bismuth nitrate $[Bi(NO_3)_3 \cdot 5H_2O]$, ferric nitrate $[Fe(NO_3)_3 \cdot 9H2O]$, glycol $[(HOCH_2)_2]$ as solvent and polyethylene glycol (PEG) $[HO(CH_2CH_2O)_nH]$ as additive. All the chemicals above were analytical grade purity. Firstly, 0.005 mol $Bi(NO_3)_3 \cdot 5H_2O$ and 0.0002 mol $Fe(NO_3)_3 \cdot 9H_2O$ were dissolved in glycol of 10 mL. PEG can be added if necessary. The solution was magnetically stirred for 3-4 hours till the solid were completely dissolved. Then dry the solution for 2days under 70 °C. Finally, after being ground in mortar for about 15 minutes, the gel powder was annealed for 30 minutes.

The structure and morphology of the $Bi_{25}FeO_{40}$ sample powders were investigated by DLMAX-2200X X-ray diffractometer (XRD) and JSM-6700F scanning electron microscope (SEM). Data for elementary analysis were got by Energy Dispersive Spectrometer (EDS). The UV-vis absorption spectra were measured on U-3010 UV-vis spectrophotometer. The photocatalytic activity of the $Bi_{25}FeO_{40}$ particles was evaluated by its degradation ratio of methyl orange $[C_{14}H_{14}N_3NaO_3S]$ (MO) under irradiation of a 500 W Xe lamp. H_2O_2 can be added as assistant if necessary. The 420 nm cut filter of the Pyrex glass cell was placed for Vis-only test.The initial concentration of MO was 20 $mg \cdot L^{-1}$, the catalyst powders was 0.2 g and the concentration of H_2O_2 was 10 g/L.

DISCUSSION

Figure 1(a) shows XRD patterns of the powders with glycol of 10ml annealed under 500 °C, 600 °C and 700 °C respectively. $Bi_{25}FeO_{40}$ with little α-Bi_2O_3 (PDF-#78-1543) was synthesized under 700 °C. α-Bi_2O_3 (PDF-#71-2274) existed when thermal annealing temperature was lower than 700 °C. It suggested that $Bi_{25}FeO_{40}$ crystal grows and the content of α-Bi_2O_3 decrease when thermal annealing temperature increases. The existence of α-Bi_2O_3 is mainly caused by nonuniform dispersion in sol and oxidation of Bi.

XRD patterns of $Bi_{25}FeO_{40}$ powders under 600 °C with different additive concentration were showed in Figure 1(b). PEG were added as dispersant. PEG have a significant influence to synthetize pure phase $Bi_{25}FeO_{40}$ and inhibit α-Bi_2O_3 growth according to diffraction pattern. It suggested that the content of α-Bi_2O_3 decrease while PEG is increasing. Pure phase of $Bi_{25}FeO_{40}$ can be fabricated with PEG of 20 g/L under 600 °C which is lower than that in former experiment. It is known that polymers like PEG has steric effect, which may on one hand improve the dispersion and stability of the particles in sol so that we can synthetize $Bi_{25}FeO_{40}$ powders under a lower annealing temperature and on the other hand prevent the aggregation of particles, and these two aspects both help reduce the particle size of the $Bi_{25}FeO_{40}$ powders so as to gain a good photocatalytic activity. It is confirmed by the photocatalytic test later. When the content of PEG increases to 30 g/L, little α-Bi_2O_3 appeared. We may deduce that steric effect of

PEG was saturated at the content of 20 g/L.

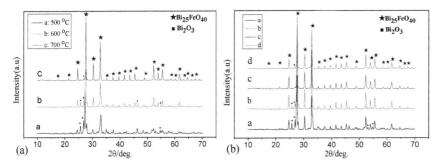

(a)

(b)

Figure 1. a) XRD patterns of $Bi_{25}FeO_{40}$ powders under different thermal annealing temperature (**a**: 500 °C **b**: 600 °C **c**: 700 °C); **b)** XRD patterns of $Bi_{25}FeO_{40}$ powders under 600 °C with different additive concentration (**a**: no additive **b**: 10 g/L PEG **c**: 20 g/L PEG **d**: 30 g/L PEG)

The morphology of the $Bi_{25}FeO_{40}$ powders annealed under 600 °C with PEG of 20 g/L was investigated by SEM. Figure 2(a) shows the as-prepared powders was petaline-like with holes, which may has a large specific surface area and is quite different from what synthesized by hydrothermal method [9,10]. The size of the $Bi_{25}FeO_{40}$ particles was about 10 μm, which is slightly smaller than the one derived from hydrothermal method [9]. It can be clearly seen that all the $Bi_{25}FeO_{40}$ particles are cubic shape from Figure 2(b). In Figure 2(c), EDS data shows that the ratio of Bi and Fe was about 23:1 which is quite close to stoichiometric ratio. As contrast, the size of the sample annealed under 700 °C without PEG is about 20 μm (Figure 2d), which confirms that PEG improve the dispersion and decrease the particle size.

151

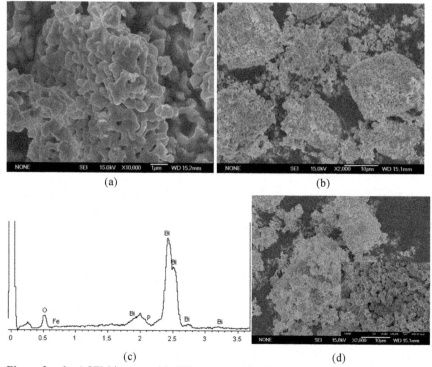

(a)   (b)

(c)   (d)

Figure 2. a,b,c) SEM images with different magnifications and EDS data of $Bi_{25}FeO_{40}$ annealed under 600 °C with PEG of 20 g/L; **d)** SEM images with magnifications of 2000 and 10000 of $Bi_{25}FeO_{40}$ powders annealed under 700 °C without PEG.

Figure 3(a) shows the absorption spectra of $Bi_{25}FeO_{40}$ powders. $Bi_{25}FeO_{40}$ powders can absorb visible light in the wavelength range of 400-550 nm. The band gap of $Bi_{25}FeO_{40}$ powders can be calculated by Kubelka-Munk formula [13], which show the relationship between $(\alpha h v)^{2/n}$ and the energy of photon hv. In this formula, n is a constant, which is equal to 4 for the indirect band gap and 1 for the direct band gap. $Bi_{25}FeO_{40}$ is direct band gap semiconductor [14], which means that n for it equals 1. The band gap of $Bi_{25}FeO_{40}$ powders could be estimated by the extrapolation of the red straight line to y=0 in the $(\alpha h v)^2 \sim hv$ plot which is shown in Figure 3(b). It indicates that Eg equals 2.2 eV, which is similar to the theoretical value of 2 eV [11] and is smller than 2.4 eV which is reported before [10], implying that $Bi_{25}FeO_{40}$ may has a better photocatalytic ability under visible-light irradiation than TiO_2, whose band gap is 3.2 eV.

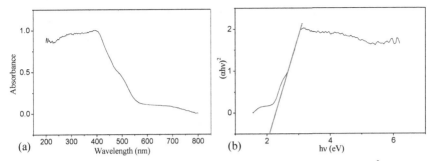

(a)

(b)

Figure 3. a) UV-vis diffuse reflectance spectrum of the $Bi_{25}FeO_{40}$; **b)** $(\alpha h\nu)^2 \sim h\nu$ plot.

Figure 4(a) shows the photocatalytic activity of $Bi_{25}FeO_{40}$ particles annealed under 600 °C with PEG of 20 g/L. MO has a high chemical stability so that it will never be degraded under UV-vis light. The degradation rate of MO was less than 2% after 3h UV-vis irradiation. However, over 43% MO was decolorized with $Bi_{25}FeO_{40}$ in 3h, which is far better than that annealed under 700 °C (8%), which is shown in Fig 4(b). Furthmore, more than 87% MO was decolorized with $Bi_{25}FeO_{40}$ assisted by H_2O_2, showing efficient photocatalytic activity of $Bi_{25}FeO_{40}$. H_2O_2 is not suitable for photocatalysis for its degradation rate of MO was less than 3% in 3h. But, H_2O_2 is beneficial for $Bi_{25}FeO_{40}$ to generate OH$^-$ which has a high photocatalytic activity [15]. It is worth mentioning that H_2O_2 had a significant influence on both samples annealed under 600 °C and 700 °C, for the 700 °C MO degradation rate increased to over 50% with the help of H_2O_2 (Figure 4b). The $Bi_{25}FeO_{40}/H_2O_2$ collaboration system remains further study. After putting a filter glass to cut off the UV light (less than 420 nm), more than 27% MO was decolorized with $Bi_{25}FeO_{40}/H_2O_2$ in 3 hours and more than 11% MO was decolorized with $Bi_{25}FeO_{40}$ only. Its photocatalytic activity under visible light is not as efficient as that under UV-vis light. The reason may be that in one hand it is known that the organic dyes like methyl orange solution has a strong absorption at 400–550 nm [12], which is the main part of the visible light absorption of $Bi_{25}FeO_{40}$ crystals according to its diffuse reflectance absorption spectra, so it will undoubtedly has influence on its photocatalytic activity under visible light. In the other hand, The particle size of $Bi_{25}FeO_{40}$ is microscale and its photocatalytic activity might be improved if nanoscale $Bi_{25}FeO_{40}$ was synthetized. Taken together, $Bi_{25}FeO_{40}$ can be considered as a novel potential photocatalyst.

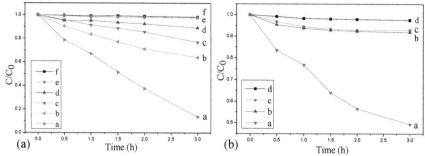

(a) Time (h) **(b)** Time (h)

Figure 4. MO degradation curve of $Bi_{25}FeO_{40}$ powder annealed under **(a)** 600 °C with PEG of 20 g/L and **(b)** 700 °C without PEG (In each diagram, **curve a:** $Bi_{25}FeO_{40}+ H_2O_2+$ UV-vis light **curve b:** $Bi_{25}FeO_{40}$ + UV-vis light **curve c:** $Bi_{25}FeO_{40}+ H_2O_2+$ vis-only light **curve d:** $Bi_{25}FeO_{40}+$ vis-only light **curve e:** H_2O_2+ UV-vis light **curve f:** only UV-vis light)

CONCLUSIONS

The sillenite phase $Bi_{25}FeO_{40}$ was fabricated via a sol-gel approach under annealing temperature of 600 °C with PEG of 20g/L. The morphology was petaline-like and its size was about 10μm. Its band gap was 2.2 eV and its 3h MO degradation rate was 27% and 87% with assistant of H_2O_2 under UV-Vis light and Vis-only light irradiation respectively.

ACKNOWLEDGEMENTS

Our work was supported by Shanghai Special Foundation of Nanotechnology under Grant No. 1052nm07300, Natural Science Foundation of Shanghai under Grant No. 08ZR1407700, Shanghai education development foundation under grant No. 08SG41, Shanghai Leading Academic Disciplines (S30107) and National Nature Science Foundation of China under Grant No. 50872080.

REFERENCES

1. A. Fujishima and K. Honda, Nature 238, 37-38 (1972).
2. S.C. Abraiiams, P.B. Jamieson and J.L. Bernstein, Journal of Chemical Physics 47, 4034-4041 (1967).
3. C.H. He and M.Y. Gu, Scripta Materialia 55 (5), 481-484 (2006).
4. C.H. He and M.Y. Gu, Scripta Materialia 54 (7), 1221-1225 (2006).
5. H.P. Zhang, M.K. Lü, S.W. Liu, Z.L. Xiu, G.J. Zhou, Y.Y. Zhou, Z.F. Qiu, A.Y. Zhang and Q. Ma, Surface and Coatings Technology 202 (20), 4930-4934 (2008).

6. X.P. Lin, F.Q. Huang, W.D. Wang and J.L. Shi, Scripta Materialia 56 (3), 189-192 (2007).
7. D.C. Craig and N.C. Stephenson, Journal of Solid State Chemistry 15 (1), 1-8 (1975).
8. S.F. Radaev, L.A. Muradyan and V.I. Simonov, Acta Crystallographica Section B 47, 1-6 (1991).
9. J.M. Li, J.Y. Song, J.G. Chen, S.W. Yu, D.G. Jin and J.R. Cheng, presented at the 2009 MRS Fall Meeting, San Francisco, 2009 (unpublished).
10. C.Y. Zhang, H.J. Sun, W. Chen, J. Zhou, B. Li and Y. B. Wang, presented at the 18th IEEE International Symposium, 2009 (unpublished).
11. C.Y. Zhang, MHA, Thesis, Wuhan University of Technology, 2010.
12. W.F. Yao, H. Wang, X.H. Xu, J.T. Zhou, X.N. Yang, Y. Zhang, S.X. Shang and M. Wang, Chemical Physics Letters 377 (5-6), 501-506 (2003).
13. D.A. Chang, P. Lin and T.Y. Tseng, J. Appl. Phys. 77, 4445-4451 (1995)
14. S.I. Stepanov, Rep. Prog. Phys. 57 (1), 39–110 (1994).
15. H.Y. Zhu, R. Jiang, Y.J. Guan, Y.Q. Fu, L. Xiao and G.M. Zeng, Separation and Purification Technology 74, 187-194 (2010).

Mater. Res. Soc. Symp. Proc. Vol. 1324 © 2011 Materials Research Society
DOI: 10.1557/opl.2011.965

Zinc Nitride Films by Reactive Sputtering of Zn in N_2-Containing Atmosphere

Nanke Jiang[1], Daniel G. Georgiev[1], Ahalapitiya H. Jayatissa[2] and Ting Wen[2]
[1]Department of Electrical Engineering and Computer Science, University of Toledo, Toledo, OH 43606, U.S.A.
[2]Department of Mechanical, Industrial and Manufacturing Engineering, University of Toledo, Toledo, OH 43606, U.S.A.

ABSTRACT

Fabrication, microstructure, chemical bonding and composition, and optical properties of zinc nitride films are investigated in this work. The films were deposited by reactive magnetron rf sputtering of zinc in N_2-Ar ambient. Based on X-ray diffraction data, the as-deposited films are polycrystalline with cubic zinc nitride structure and (400) preferred orientation. Well defined Zn-N, N-N, as well as Zn-O and H-O bonding configurations are revealed by X-ray photoelectron spectroscopy data. The as-deposited films are found to be almost-stoichiometric and contain only a small fraction of oxygen. A direct band gap of 1.5 eV is obtained by using the photon energy dependence of the optical absorption of the films. This result is confirmed independently by spectroscopic ellipsometry.

INTRODUCTION

Zinc nitride (Zn_3N_2) is a promising material for electronic and optoelectronic applications [1-4], as well as photovoltaic and sensor applications, offering additional advantages such as environment-friendly processing and potentially low fabrication cost. Zn_3N_2 powder was first synthesized by Juza and Hahn [5] in 1940. Zinc nitride films were first prepared by Kuriyama et al. [6] in 1993. Since then, different methods have been used to prepare zinc nitride films: reactive sputtering [2,7-9], chemical vapor deposition [1], and electrochemical processes [10]. However, there are still many challenges in the preparation of high-quality zinc nitride. Reproducibility, film stability, chemical composition control, and undesired oxidation of films are some of the problems with this material.

Zinc nitride remains a relatively new material and its physical properties are not well studied or understood. The values of some important parameters are not certain, and some vary over a large range, depending on either the structure/morphology, or the preparation method or techniques for characterization. This is especially true for the energy band gap, the data for which vary from 1.06 to 3.4 eV with a direct band gap or 2.11 to 2.81 eV with indirect band gap [11].

In this work, which is in progress, zinc nitride films were fabricated by reactive sputtering of Zn in Ar+N_2 atmosphere. The nitrogen concentrations in the sputtering gas mixture and the substrate temperatures are the deposition parameters which were varied. The microstructure, morphology and optical properties of the films were studied, and the effects of the deposition conditions on the crystallinity and the energy band gap were investigated. Information about the chemical bonding configurations within the films and at their surface was obtained as well. We discuss data on only three samples, representing parameter variation of two different kinds in this paper. However, many of the statements made in this paper appear supported by a larger set of samples that we are currently working on.

EXPERIMENTAL DETAILS

The zinc nitride films were deposited on both single-crystal, (100)-orientation silicon and glass substrates using a magnetron sputtering system (Torr International, Inc.) with a 3-inch diameter Zn target (purity 99.995%, Kurt J. Lesker). The substrates were mounted on a rotating holder at a distance of 10 cm from the target in a sputter-up geometry. The base pressure was in the low 10^{-6} Torr range, and the deposition pressure was fixed at 8 mTorr range by maintaining the total gas flow at 40 sccm. All films were sputtered at a rf power of 85 W in a mixture nitrogen (99.9993%, Airgas, Inc.) and argon (99.999%, Airgas, Inc.). Prior to the film deposition, a pre-sputtering step was performed for 5 min at a rf power of 80 W in Ar gas atmosphere. The targeted final thickness of the films was 500 nm and the deposition rate was monitored by a quartz crystal thickness monitor. The thickness was later verified by profilometry.

The X-ray diffraction (XRD) work was done on a PANalytical X'Pert Pro MPD and Scintag XDS-2000 X-ray Powder Diffractometer with Cu Kα radiation source in a 2θ range of 30° to 60°. X-ray photoelectron spectroscopy (XPS) was performed using a Perkin-Elmer 5500 X-ray photoelectron spectrometer with Al Kα radiation. Scanning electron microscopy (SEM) and energy dispersive X-ray spectroscopy (EDS) were generated using a Hitachi S-4800 II Field Emission Scanning Electron Microscope (FESEM). The optical transmittance and reflectance of the films was measured with a Lambda 1050 UV/VIS/NIR spectrometer in wavelength range from 350 nm to 2000 nm at room temperature (RT). Spectroscopic Ellipsometry (SE) was conducted using an M-2000FI^{TM} Spectroscopic Ellipsometer (J. A. Woollam Co., Inc.).

RESULTS AND DISCUSSION

Structural Properties

XRD patterns from films deposited on glass substrate at three different substrate temperature/nitrogen concentration combinations: 200 °C/80%, 200 °C/20% and RT/80%, are shown in Figure 1. As it can be seen in figure, with the increase of the N_2 content in the sputtering gas, the films' crystallinity improves. At a N_2/N_2+Ar ratio of 20%, only a very weak XRD diffraction peak located at 36.8°, associated with the (400) plane of cubic anti-bixbyite structure of zinc nitride [6,12-14], is detected. When N_2 content increases to 80% of the total sputtering gas mixture, the (400) plane zinc nitride peak becomes much sharper. The full width at half maximum (FWHM) value decreases from 0.96° at 20% to 0.28° at 80%, which means film crystallinity is improved noticeably. Also, the crystallinity improves with the increase of the substrate temperature, as indicated by the FWHMs of the (400) plane zinc nitride peak, 0.35° at RT and 0.28° at 200 °C.

158

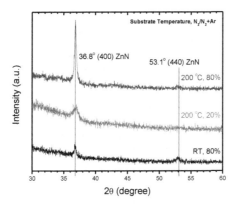

Figure 1. XRD patterns for samples deposited at three different substrate temperature/nitrogen concentration combinations: 200 °C/80%, 200 °C/20% and RT/80%. The thickness of all films is 500 nm and the substrate is glass. The data are not normalized and are not corrected for any contribution from the substrate. The step size is 0.0167 deg.

Chemical Bonding and Composition

XPS , SEM and EDS were used to study the chemical bonding states, surface morphology and chemical composition on the zinc nitride film prepared at the $N_2/(N_2+Ar)$ ratio of 80% and substrate temperature of 200 °C.

XPS spectra were taken from both as-deposited film surfaces and surfaces of film after 10 min Ar-ion sputtering (using argon-ion energy of 1.5 keV and a current of 20 mA). Figure 2 shows the N1s and O1s spectra of zinc nitride film both at the as-deposited surface and the

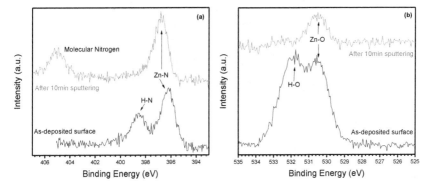

Figure 2. (a) XPS N1s and (b) O1s high-resolution spectra of zinc nitride film deposited at a substrate temperature of 200 °C and N_2/N_2+Ar ratio of 80%.

159

surface after 10 min sputtering treatment. In the surface spectrum of Figure 2(a), two peaks, located at 396.27 eV and 398.57 eV, are observed. They correspond to N-Zn bonding, and N-H bonding configurations [7]. In Figure 2(b), at the surface, the lineshape consists of a H-O peak at 531.95 eV and a Zn-O peak at 530.39 eV [15]. It has been reported that the surface of such films can be water-contaminated [7] as evidenced by well defined H-N and H-O bonding. After argon-ion sputtering(i.e., after removing a thin surface layer), the peaks are shifted to lower binding energy which can be attributed to partial de-nitrification [7] and therefore the film becomes more metallic and likely more conducting. However, it is also worth noticing that: first, the water-contamination is removed, showed by the dramatic decrease in the H-O and H-N features; second, the intensity of the Zn-O bonding is significantly reduced while the intensity of Zn-N bonding increases; third, a molecular nitrogen bonding peak appears (i.e., N-N bonding), located at 405.03 eV [16], which may explain the excess nitrogen content, that is sometimes observed in EDS measurements (not shown or discussed in this text).

SEM and EDS characterization was performed on zinc nitride films, deposited on Si substrates, under the same deposition conditions. A voltage of 5 kV and a working distance of 15 mm were used. An SEM image representing the typical film surface morphology is shown in Figure 3, and Table I shows the EDS compositional analysis results for the same film.

It can be seen from Figure 3 that the grains pack compactly and the grain size is around 15 nm, which is in agreement with an independent mean grain size value evaluation, using the Scherrer's formula and the XRD data. According to Table I, the film is almost stoichiometric (the concentration measurement error for nitrogen is about 5%, and better for the higher-Z elements).

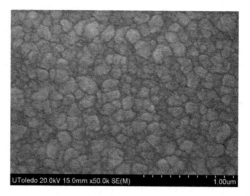

UToledo 20.0kV 15.0mm x50.0k SE(M)   1.00um

Figure 3. SEM image of zinc nitride film deposited at a substrate temperature of 200 °C and N_2/N_2+Ar ratio of 80% .

Also, based on the EDS results, only small amount of oxygen is present in the as-deposited film. This noticeable fraction of oxygen is likely due to water-contamination as mentioned in the XPS part. At this point, our data cannot address the issue of how the nitrogen and the oxygen are distributed in the film and what faction of them participates in Zn-N or Zn-O bonds.

Table I. EDS results for zinc nitride film deposited at a substrate temperature of 200 °C and N_2/N_2+Ar ratio of 80%. The acceleration voltage was 5kV and no Si signal (from the substrate) was observed at this voltage.

Element	N	Zn	O
Atomic %	36.5	58.5	5.0

Optical Properties

Optical transmittance spectra for three different substrate temperature/nitrogen concentration combinations: 200 °C/80%, 200 °C/20% and RT/80%, are presented in Figure 4. Films show much higher transmittance in the near infrared range than in the visible range. This agrees with the results reported by T. Yang *et al.* [12]. In addition, it needs to be noticed that the transmittance decreases with the increase in the nitrogen ratio, while the substrate temperature impact is only minimal.

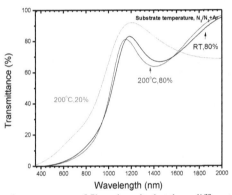

Figure 4. Optical transmittance spectra of films deposited at three different substrate temperature/nitrogen concentration combinations: 200 °C/80%, 200 °C/20% and RT/80%.

Because the reflectance of the films is dependent on the wavelength, we needed to take this into account when calculating the absorption coefficient. Then, by using the Tauc model (plotting $(h\nu\alpha)^2$ versus energy, for direct electron transition) and the Davis-Mott model (plotting $(h\nu\alpha)^{1/2}$ versus energy, for indirect electron transition) [11], the band gap of the film deposited at N_2/N_2+Ar ratio of 80% and substrate temperature of 200 °C was determined to be 1.49 eV and of a direct type. This was also confirmed by our first spectroscopic ellipsometry data (not shown).

CONCLUSIONS

Zinc nitride thin films were deposited by reactive rf sputtering. Structural, chemical bonding and compositional information was obtained, and the optical properties of the films were examined. The films are polycrystalline with a preferred orientation corresponding to the (400) plane of the cubic anti-bixbyite structure of zinc nitride, as revealed by the XRD measurements. The crystallinity of the films increases with the increasing of either the nitrogen ratio or the substrate temperature. According to the XPS and EDS results, well defined Zn-N and Zn-O bonding indicate that zinc nitride, combined with only a small amount of zinc oxide, were formed during deposition. The surface of film was water contaminated and had a higher oxidation level. The optical properties of the films were investigated independently by optical transmittance measurement and spectroscopic ellipsometry. The band gap was found to be about 1.5 eV of a direct type.

ACKNOWLEDGMENTS

This work was supported by the National Science Foundation (CMMI, award # 0928440). The XPS work was performed at the Center for SSIM, Wayne State University, Detroit, Michigan, with the help of Dr. Erik McCullen. We would like to also thank Dr. Robert Collins and his group from the University of Toledo's Department of Physics and Astronomy, for help with ellipsometry measurements.

REFERENCES

1. D. Wang, Y.C. Liu, R. Mu, J.Y. Zhang, Y.M. Lu, D.Z. Shen, and X.W. Fan, J. Phys.: Condens. Matter 16, 4653 - 4642 (2004).
2. V. Kambilafka, P. Voulgaropoulou, S. Dounis, E. Iliopoulos, M. Androulidaki, K. Tsagaraki, V. Saly, M. Ruzinsky, P. Prokein, and E. Aperathitis, Thin Solid Films 515, 8573 - 8576 (2007).
3. T. Oshima, and S. Fujita, Jpn. J. Appl. Phys., Part 1 45, 8653 (2006).
4. N. Pereira, L.C. Klein, and G.G. Amatucci, J. Electrochem. Soc. 149, A262 (2002).
5. H. Juza, and Z. Hahn, Anorg. Allg. Chem. 224, 125 (1940).
6. K. Kuriyama, Y. Takahashi, and F. Sunohara, Phys. Rev. B 48, 2781 (1993).
7. M. Futsuhara, K. Yoshioka, and O. Takai, Thin Solid Films 322, 274 – 281 (1998).
8. V. Kambilafka, P. Voulgaropoulou, S. Dounis, E. Iliopoulos, M. Androulidaki, V. Saly, M. Ruzinsky, and E. Aperathitis, Superlattices and Microstructures 42, 55 – 61 (2007).
9. C. Wang, Z.G. Ji, K. Liu, Y. Xiang, Z.Z. Ye, J. Cryst. Growth 259, 279 – 281 (2003).
10. K. Toyoura, H. Tsujimura, T. Goto, K. Hachiya, R. Hagiwara, and Y. Ito, Thin Solid Films 492, 88 – 92 (2005).
11. W. S Khan, and C. Cao, J. Cryst. Growth 312, 1838 – 1843 (2010).
12. T.L. Yang, Z.S. Zhang, Y.H. Li, M.S. Lv, S.M. Song, Z.C. Wu, J.C. Yan, and S.H. Han, Appl. Surf. Sci. 255, 3544 – 3547 (2009).
13. O. Takai, M. Futsuhara, G. Shimizu, C.P. Lungu, and J. Nozue, Thin Solid Films 318, 117 – 119 (1998).
14. W. Du, F.J. Zong, H.L. Ma, J. Ma, M. Zhang, X.J. Feng, H. Li, Z.G. Zhang, and P. Zhao, Cryst. Res. and Technol. 41, 889 (2006).
15. M. Futsuhara, K. Yoshioka, and O. Takai, Thin Solid Films 317, 322 – 325 (1998).
16. G.Z. Xing, D.D. Wang, B. Yao, L.F.N. Ah Qune, T. Yang, Q. He, J.H. Yang, and L.L. Yang, J. Appl. Phys., 108, 083710 (2010).

AUTHOR INDEX

163

SUBJECT INDEX

Printed in the United States
by Baker & Taylor Publisher Services